David Rattlehead

The Complete Practical Distiller

Comprising the most perfect and exact theoretical and practical description of the

art of distillation and rectificiation

David Rattlehead

The Complete Practical Distiller
Comprising the most perfect and exact theoretical and practical description of the art of distillation and rectificiation

ISBN/EAN: 9783744750691

Printed in Europe, USA, Canada, Australia, Japan

Cover: Foto ©berggeist007 / pixelio.de

More available books at **www.hansebooks.com**

THE COMPLETE PRACTICAL DISTILLER:

COMPRISING

THE MOST PERFECT AND EXACT THEORETICAL AND PRACTICAL DESCRIPTION

OF THE

ART OF DISTILLATION AND RECTIFICATION;

INCLUDING ALL OF

THE MOST RECENT IMPROVEMENTS IN DISTILLING APPARATUS; INSTRUCTIONS FOR PREPARING SPIRITS FROM THE NUMEROUS VEGETABLES, FRUITS, ETC.

DIRECTIONS FOR THE DISTILLATION AND PREPARATION OF

ALL KINDS OF BRANDIES AND OTHER SPIRITS, SPIRITUOUS AND OTHER COMPOUNDS, ETC., ETC.

BY

M. LA FAYETTE BYRN, M.D.,

GRADUATE OF THE UNIVERSITY OF THE CITY OF NEW YORK.

EIGHTH EDITION.

TO WHICH ARE ADDED,

PRACTICAL DIRECTIONS FOR DISTILLING.

FROM THE FRENCH OF TH. FLING, BREWER AND DISTILLER.

PHILADELPHIA:
HENRY CAREY BAIRD & CO.,
INDUSTRIAL PUBLISHERS,
810 WALNUT STREET.
1880.

PREFACE TO THE EIGHTH EDITION.

IN presenting to the public a new and improved edition of THE COMPLETE PRACTICAL DISTILLER, the publisher desires to express an acknowledgment of his appreciation of the success with which the book has met in the past, and of its growing popularity at the present time.

He has added to it M. Flinz's PRACTICAL DIRECTIONS FOR DISTILLING, which has recently appeared in Paris as a separate and distinct publication. This, he confidently believes, will be found direct and practical, and will add greatly to the value and usefulness of a book which has already received so many and such substantial evidences of favor.

H. C. B.

PHILADELPHIA, March 15, 1870.

PREFACE.

For a long time the public have been in want of a work on the art of Distillation and Rectification, couched in such language that every one could appreciate it; and of such size and value that the price, and the time required to read it, would prove the least objectionable. From the best information I can gain, no work has appeared on this subject for many years. Owing to this fact, most of the improvements which have been made in the art have proved of little use to the larger class; and thus things have almost remained stationary with regard to this very important matter, particularly in this country; which is, indeed, greatly to be lamented, as we are in possession of every thing, in the way of fruits, vegetables, etc., which have hitherto been used in distillation.

I trust that in the following pages the reader will find every thing that the present state of science calls for, and that the suggestions may prove of great practical advantage; which I think they will do, as every thing is given in the shortest and plainest manner. It is almost needless to say that I have consulted every authority that I could find, for the purpose of making this a complete work; they are, however, too numerous to mention here, and would be, moreover, of no benefit to the reader. With these few prefatory remarks, the book is submitted to the public.

M. La Fayette Byrn, M. D.

CONTENTS

	PAGE
Description of a Distillery	9
Some Directions to the Distiller	11
Of Distillation, and the Apparatuses made use of	17
Continuous Distillation	26
Mode of Working the Apparatus	32, 39, 43
Apparatus used principally in American and English Distilleries	44
Instrument to prevent Inequality of Heat in Distillation	59
Of the Process of Malting, etc.	63
French Method	79
English Method	81
Fermentation	84
Rectification	89
Common Process of Malt Distilling	91
French Process of Distilling and Preparing Brandy	93
Method of Preventing the Deterioration of Brandies	95
Malt Whisky	96
Process for Making Dutch Geneva	98
Process for Brewing Hollands Gin	101
Process for Rectification into Hollands Gin	103
Distillation of Common Gin	106
Spirit of Potatoes	106
Apparatus made use of in the Distillation of Potato Spirit	107
Reduction of the Potatoes	112

CONTENTS.

	PAGE
Mashing of Potatoes	114
Rasping Potatoes	116
Separation of the Fecula	116
Draining	118
Arrack, or Spirits of Rice	124
Spirits of Beet-Roots	127
The Beet Rasp	128
Kirsch-Wasser, or Spirits of Cherries	133
Of some of the Products of this Country which afford Spirits by Distillation	135
Cider Spirits, or Apple Brandy	135
Peach Brandy	136
Of the Preparation and Distillation of Rum	137
Process made use of in Great Britain and Ireland for Fermenting and Distilling Molasses	140
Raisin Spirits	143
Flavoring and Coloring of Spirits	143
Process for Making Rum Shrub	144
Process for Making Brandy Shrub	145
Elder Juice	145
Method of Making Cherry Brandy	146
Eau de Luce	147
Irish Usquebaugh	148
Process of Making Nectar	149
Imperial Ratafia	149
Method of Making Lovage Cordial	150
Process of Making Citron Cordial	150
Cinnamon Cordial	151
French Noyau	151
Peppermint Cordial	152
Process of Making Aniseed Cordial	152
Method of Making Caraway Cordial	153
French Vinegar	153
Method of Making English Vinegar	154

CONTENTS.

	PAGE
Some General Directions for the Distillation of Simple Waters, etc.	155
Of the Stills used for Simple Waters	156
Cinnamon Water	158
Peppermint Water	158
Damask-Rose Water	158
Orange-Flower Water	158
Orange Wine	159
Simple Lavender Water	159
Compound Lavender Water	160
Hungary Water	160
Some General Directions for the Distillation of Spirituous Waters	161
Jessamine Water	162
Eau de Beauté	162
Some Remarks on the Uses of Feints, and their General Character	163
Rules for Determining the Relative Value and Strength of Spirits	164
Observations on Distillations of a Special Character, and on the Selection of Apparatus most useful	165
Remarks on an Instrument intended for Testing Wines	184
Some General Directions for the Preparation of various Cordials, Compounds, etc.	187
On some of the Plans resorted to for the purpose of Adulterating Brandy	188
Process for Making Lime Water	191
Process of Making Sulphuric Ether	191
Instructions for Making Infusions, Spirituous Tinctures, etc.	194
Tonic and Alterative Cordial	195
Aromatic Bitters	196
Process for Making a Diuretic and Stomachic Compound	196
Process for Making Tincture of Musk	197

APPENDIX.

PRACTICAL DIRECTIONS FOR DISTILLING. FROM THE FRENCH OF TH. FLINZ, BREWER AND DISTILLER.

PART FIRST.

PRELIMINARY OBSERVATIONS.

	PAGE
I. Maceration	200
II. Fermentation	202
III. Distillation	204
IV. Rectification	205

PART SECOND.

SPECIAL OBSERVATIONS.

I. Buildings	207
II. Utensils	208
III. Maceration	208
IV. Fermentation, Distillation, Rectification	209
V. Yeast	210
VI. Malt	211
VII. Preservation of Spirituous Liquors	213
VIII. Raw Materials	214
Index	215

THE COMPLETE PRACTICAL DISTILLER.

DESCRIPTION OF A DISTILLERY.

When the establishment of a distillery on a grand scale is undertaken, it is incumbent on those concerned to make every preparation necessary to facilitate their labours, insure the preservation of their materials, preserve their products, and employ as few hands as possible. The space destined for a distillery should of course be large. It should contain a plentiful spring, excellent vaults, store-houses, &c. A situation near a stream of water is, of all others, the most preferable, if in the country; but by whatever means water may be obtained, it will be necessary to be secured against the possibility of a failure at any time.

The cellar should be considered as the magazine in which all the wine, previous to its distillation, should be deposited; and ought to occupy the same space under ground as the distillery above it. It has been observed that the best and most perfect cellar is that where the thermometer is always between 55° and 65° of heat by the scale of Fahrenheit. The further the temperature of this part deviates from this standard, the worse it is. If a cellar has not a sufficient depth, it is necessary to dig it deeper; if too much exposed to the air, surround it with walls; increase the doors, and diminish the air-holes; stop up those that are not well placed, and open fresh ones that will introduce a new current of air.

A cellar ought to be at least about sixteen feet in depth, the roof twelve or fourteen feet high, and the whole bottom covered with some four feet of earth. The entrance should always be within two doors, one of which should be at the top of the stairs, and the other at the bottom; and this is equal to a gallery. If the entrance should look toward the south, it is necessary to change it, and carry it to the north. Cellars whose entrances are toward the south or the west are not as they should be; every one must see the reason of this. In proportion as the heat of the atmosphere after winter increases eight or ten degrees, a certain number of the air-holes must be closed, because the air of a cellar always endeavours to put itself in equilibrium with that of the atmosphere. On the contrary, during the summer it is proper to admit the external air to a certain point, to diminish the heat of the cellar. Here, however, some restriction is necessary: if the external air is of 55°, then the air-holes must be closed. Prudent conduct with respect to the air-holes will preserve the wine, and prevent its being impaired while in the casks.

A good cellar for wine, spirits, or beer should be at a proper distance from the passage of carts, carriages, and all manner of vehicles; and also from shops or forges of workmen who are continually in the use of the hammer and anvil. Their blows affect the vessels, as well as the fluids they contain; they also facilitate the disengagement of the carbonic acid gas, the first connection of bodies; the lees combine with the wine, insensible fermentation is augmented, and the liquor more promptly decomposed.

A cellar cannot be too dry; humidity undermines the tuns, moulds and rots the hoops till they burst, and the wine is lost. Besides this, humidity penetrates the casks insensibly, and at length communicates a mouldy taste to the liquor. Experience has proved in France that wine preserved in vast tuns, built into the stone walls of good cellars, increases in spirit every year. These tuns are not subject to running, like the common casks; and also contribute very much in point of economy, and in the end are less expensive than wood. For one apparatus, the space appropriated to a distillery, properly speaking, should not be less than from forty to fifty feet by fifteen or twenty; but this is only to be understood of distilleries of wine or spirits. A large yard or court is also necessary to a distillery.

SOME DIRECTIONS TO THE DISTILLER.

THE average gravity of worts brewed from a mixture of malt and barley is, in all, from 100 to 120 pounds of saccharine matter per barrel. But part of this gravity is made up from a mixture called *lob*, which is a powerful and strong saccharine, made from barley and malt flour, and added to the brewing of the common worts. This mixture, although so high in gravity, is yet generally well fermented, being cut down so low as from 6 to 2 pounds on Dicas's instrument, (given further on.) This attenuation is accomplished generally in the space of from 10 to 20 days at most. When perfectly fine, it is put into the wash-still, and distilled into low wines. These are afterward put into the low wine still, and made into *spirits* and *feints*. The mere working of these stills is a simple mechanical process, to perform which, from their great size, there is plenty of time.

The average charge of a wash-still is from 10,000 to 20,000 gallons of wash at once, and the charge of the low wine still is the produce of the wash from the wash-still. From this it will be seen that the particular still requisite in conducting a distillery to advantage, relates to the brewing of strong worts, and to the proper fermenting of them, a sort of knowledge which has absolutely become a science in the hands of those who possess it.

When the still is charged with goods for distilling, and luted, then make the fire under it, which should be of coals, if they can be obtained, because their heat is most durable, and wood fires are subject to both extremes, of too much and too little heat, which are prejudicial and hazardous. Let the fire be pretty moderate at first; then increased by degrees, and now and then stirred up with the poker; and by laying the hand upon the body of the still, as the fire gains strength in the stove or furnace under the still, you will by moderate degrees carry it up to the still-head. When this becomes warm or hot, a damp is to be prepared to check or lessen the violence of the fire. Special care must be taken that no manner of grease, tallow, soap, or any other such like unctuous matter, get or fall into the tubs, rundlets, or cans, because they quite take off all manner of proof of the goods; and although the strength be very high, yet they will apparently fall as flat as water, and then their strength can only be ascertained by the hydrometer. Lighted candles, torches, paper, or other combustible matters, should never be brought near the still or any vessel where the goods are contained, which are subject to take fire upon very slight occasions.

But should an accident take place, get immediately a woollen blanket or rug, drenched in water, and cast upon the flame, which will extinguish it by excluding the air. Some persons, after the still is charged, make a luting or paste, made half of Spanish whiting and the other of rye-meal, bean-meal, or wheat-flour, well mixed together, and made with water of the consistence of an ordinary paste for baking; and having put on the still-head, work and make it pliable, and spread it upon the junctures of the body and head of the still, to keep in the goods from boiling over. Reserve a piece of the paste, lest the luting should crack or break out, which is very dangerous. It is a custom among some gentlemen of the trade to put one-third or one-fourth part of proof molasses-brandy proportionally to what rum they dispose of, which cannot be distinguished but by an extraordinary palate, and does not at all lessen the body or proof of the goods, but makes them something cheaper. To recover or amend any common waters, or genevas, will take such a quantity of proof or double goods of the same kind or denomination to the other as the price will bear, or will answer the intentions, by such composition or mixture.

If by putting proof and weak goods together, the colour or face of the goods be spoiled, which before their being mixed together were fine, as it frequently happens, they must be cleaned or fined, as when newly distilled. Some persons throw in about a pound of alabaster powder into their mixed goods, to stop up the porous parts of the flannel sleeve, which fines them immediately.

To recover any goods to a better body or strength, when too low or weak, or fine cordial waters, a proper quantity must be put, by little and little at a time, of spirits of wine to the goods, mixing or stirring them well together. They may be perfectly restored to the desired proof with little or no loss, because the spirits of wine stand at about the same price with the cordials, and cost less than some of the brandies. If, by chance or accident, any goods happen to be spoiled in their complexion, especially genevas, which may be turned as black as ink even by an iron nail dropping into the cask, they must be distilled over again, by putting in half the quantity of the ingredients as usual; and they will come perfectly fine as rock-water from the still, and must be dulcified according, just as they were at their first being made. But the goods, notwithstanding the misfortune they met with, will be much better than they were before; for by every distillation they are weakened near 1 in 20, though improved in goodness, as before observed.

Distillers, when drawing off and making up their distilled goods, should be often trying them in a glass or phial; and when the bead or proof immediately falls down, and does not continue a pretty space upon the surface, then they should take away the can of goods, and substitute another vessel to receive the feints, which, if suffered to run among the rest, would cause a disagreeable relish, and be longer in fining down; whereas, the feints being kept separate, the goods will be clean and well tasted when made up with liquor to their due quantity. When the still is first charged, some persons add about 6 ounces of bay-salt to every 10 gallons of spirits, and so proportionably, whereby the goods will cleanse themselves, and separate from their phlegmatic parts. Some are also in the habit of using a handful of grains of paradise, to make the goods feel hot upon the palate, as if they bore a better body; yet this should never be done, as it conduces nothing toward the advancement of the proof.

After all the goods have come off, if designed for double goods, they must be made up to their first quality with liquor. For instance, if a still is charged with 3 gallons of proof spirits, they will yield in distillation about 2 gallons without feints; which deficiency of 1 gallon must be made up with liquor (and sugar used in dulcifying) to their determined quantity. To single or common goods must be added, over and above the prescribed quantity in compounding double goods, one and a half part more of liquor, (viz. one gallon and a half,) to dilute it for single or common goods.

When goods are to be dulcified, you must never put your dissolved sugar among your new distillation till the dulcifying matter becomes perfectly cold; for if mixed hot with the goods, it would cause some of the spirits to exhale, and render the whole more foul and phlegmatic than otherwise. To fine any goods speedily for immediate use or sale, (especially white or pale goods,) add about 2 drachms of crude alum, finely powdered, to 3 gallons of goods; rummage them well, and the residue will immediately become clear and transparent. It must also be observed, that what is called the Hippocrates bag, or flannel sleeve, is very necessary for a distiller or brandy-merchant, as by the use of this all bottoms of casks, though ever so thick and feculent, by putting into this bag to filter, become presently clear—the porous parts of said bag being soon filled with grosser matter, and the thin or liquid element runs clear from the bag, and is as good as any of the rest. Also, any foul goods or liquor may be presently made clear and fine, by putting some alabaster, powdered, into the liquor, or sprinkling the same on the bag to stop its pores, by which they presently become or run clear, leaving nothing but the sediment or gross matter in the bag; nor does the liquor contract the least ill flavour from the alabaster powder.

The said bag is made of a yard of flannel, not over fine or close wrought, laid sloping, so as to have the bottom of it very narrow, well sewed up the side, and the upper part of the bag folded about a broad wooden hoop, and well fastened to it; then boring the hoop in three or four places, it may be suspended by a cord. But the bottoms of fine goods, which are much more valuable, must be filtered or put through blotting-paper, folded in four parts, one part or leaf to be opened funnel-wise, and made capable to receive what it will hold of the bottoms; this being put into the upper part of a large tin funnel, will filter off all the goods from the sediment.

OF DISTILLATION, AND THE APPARATUSES MADE USE OF.

THE apparatus for distilling, upon which many improvements in France are founded, is that of M. Adam. In a furnace, situated in one corner of the distillery, is placed a still built into the masonry. The head is in the form of a dome, solidly fixed with the cucurbit. From the centre of this dome a tube ascends, as thick as a man's arm; and this runs into the first vessel, placed on one side of the still, which is fixed upon strong joists.

From this vessel issues a second tube, similar to the first, but in the form of an arch, which enters into another vessel, also resembling the first, which communicates with a third in the same manner. In this apparatus, thus simplified, there are several points to be considered: In the first place, all the vessels fixed upon the joists are made in the form of an egg, and have their two ends placed vertically. Secondly, that the entering tubes, viz. those which proceed from the still to the first egg, and from the first to the second, &c. have their extremities in the bottom of each egg, and there form something like the head of a garden or watering pot, pierced with several holes. Thirdly, the last of these eggs, when there are but three, and sometimes the two last, when there are four, are furnished with a cooler in their upper part; and this is always filled with water while the distillation is going on. These vessels, with their refrigerators, are called condensers.

Every distiller does not use condensers; the majority

look upon them as useless when they only wish to obtain three-six. However, they have all the rest of the apparatus complete; and as these eggs communicate one with another, and each separately with the first worm, they may be used as condensers at pleasure, it is only necessary to turn or stop one of the cocks.

At the extremities of these eggs a large tub is placed, the interior of which contains a large worm constructed of tin, which plunges into the wine instead of water, and is hermetically sealed. This first worm communicates with a second longer than itself, and enters a large tub placed under the first, which is entirely full of water.

On one side, and under this lower tub, a large space is dug in the earth and built round with stone, which the French distillers call a *tampot;* this serves as a magazine for their wine previous to distillation, which may be pumped into the upper tub. All the eggs, as well as the still, communicate with the upper tub through tubes placed between the lower part of the eggs and the still; there are, besides, lateral tubes which run from the upper part of the eggs to the orifice of the worm in the upper tub. There are other tubes proceeding from the upper part of each of the vessels, even from the still, which enter a small worm immersed in a little tub upon the furnace, by the side of the still. The mechanism of the distillation is no less curious than the apparatus.

Explanation of the Egg-Plate.—A is the furnace on which the still B is built; of this the dome or head only is to be seen; the punctuated lines indicate the form masked by the building. C is the tube, furnished with a cock on the outside of the furnace, communicating with

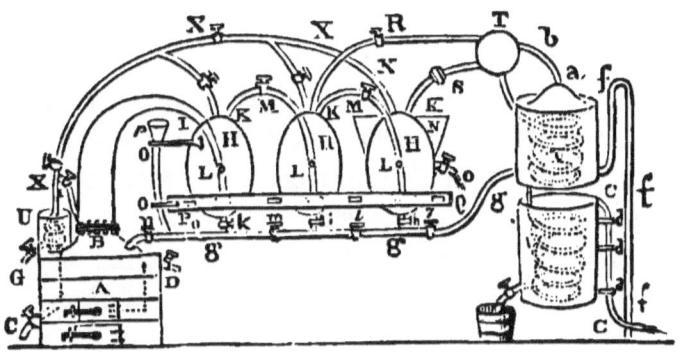

Fig. 1

the bottom of the still, for the purpose of discharging the alembic and the eggs. The small tube D, also provided with a cock, serves to point out when the still is full within two-thirds of its height. The little tube E also proceeds from the head of the still, with its cock, which communicates with the long tube x x x x, which runs from the last egg—that is to say, from that at the greatest distance from the still—and communicates with the little worm which is plunged in the little tub F, placed under the furnace to prove the vapours contained in each of the distillatory vases. This little worm has the cock G at its lower orifice. H, H, H are a series of distillatory vessels or condensers, in the shape of eggs, solidly fixed upon the timber-work P Q, and in succession with each other on the side of the still.

This plate represents only three eggs, though the number may be augmented at pleasure. It was the opinion of M. Adam that the greater the number of eggs, the better the rectification would be carried on. The still communicates with the first egg by the tube I, which rises from the centre of the head or dome, and descends to the bottom of the egg, where it enlarges into the form of the rose of a garden watering-pot, pierced with a number of holes. It must be understood that this tube is soldered to the egg at its entrance, to prevent any other issue of the vapours but by the way intended.

The first egg communicates with the second, this with the third, and so on to the last, by means of the tube M, which is soldered to the first egg at the point K, and proceeds to the bottom of the following, where it enlarges in the form of a watering-pot, as in the first. The last egg is furnished with the cooler N, by means of which the superior part of the egg, where the vapours are collected, is encircled with water to commence the refrigeration. This cooler is supplied with a cock O, to let out the water when it gets too warm. Every condenser is furnished with a cock like this, or otherwise their upper parts are plunged into the common tub full of water.

This tub or bag, often made of copper, has the form of a parallelepiped. The tube R communicates from the second egg with the worm, which is generally used with two eggs, sufficient to obtain brandy at 18°, when they close the cock M, which communicates with the second and third egg, and they open the cock R to establish the communication with the worm. The pipe S communicates between the third egg and the worm. When three eggs are used, they operate as just indicated; they open the cocks M and S, and stop the cock R. The same proceeding is observed when the greatest number of eggs are employed.

Each of these has a tube that communicates with the

worm, and all these are soldered to the spherical T, in which the vapours from each egg are deposited, to be conveyed from thence into the worm in the tub U. U is a tub, hermetically closed, which contains the principal worm; this is full of wine, heated by the passage of the hot vapours from the last. It is also surmounted with the dome a, from which proceeds the pipe b, that serves to contain the alcoholic vapours that escape from the tube last mentioned, from the vessel T, or from any of the eggs or still, to convey them thence into the worm. J is a large tub under the first, and which encloses the second worm, but is much longer than the other.

It is full of water, always kept cold; but disgorges itself through the pipe c on the outside of the vessel, against which it is supported by the three iron bars d, d, d. It has not been thought necessary to represent the stone cavity used as a storehouse for the wines designed for distillation, which wines may be raised into the tub U by means of a pump managed by one man; the conducting pipe of this, marked fff, discharges itself near the bottom of the tub U.

ggg is the pipe of communication belonging to the still and the eggs; h, i, k are cocks to establish or intercept the communication of the eggs with the conducting pipe g; l, l, m, n are cocks for continuing or interrupting the communication between each egg and the still, to discharge it, or with the condensing vessel, for the purpose of filling it; $o\ o\ o$ is the pipe through which the brandy or the feints are conveyed by means of the tun p, when they wish to charge the still or the eggs. It is soldered to the pipe g, into which it discharges itself, and is consolidated with the rest of the apparatus by two iron bars, one of which is nailed to the timber-work P Q, while the other is attached to the first egg. This pipe is called *corne d'abondance*, or horn of plenty. All the apparatus of the French distillers that have been encouraged by patents have been constructed according to the principles of this now described, or those analogous to them.

In the working of the still just described, they first close the lower cocks that communicate with the grand tube connected with the egg. They open those of the conducting tube; then the wine contained in the tun escapes and settles in the still. During this time a labourer pumps, to replace the wine in the tun that has escaped by the pipe. They know that the still is sufficiently charged when the wine flows through the little cock adapted to it. The globules are compelled to traverse the liquid to ascend to the upper part of the egg; but it is necessary to observe that the vapours that issue from the still are not purely alcoholic, but mixed with many watery particles.

In visiting the vacant part of the egg, the watery part mixes with the wine, with which it has much affinity, while the spurious parts, accumulating in the upper part of the first egg, pass from that into the second and third, and after having traversed them all, settle in the upper worm, where they condense, and finish the cooling in the second worm.

The liquor comes out cold from the lower orifice of the second worm, and is received into the vessel destined to that purpose. The vapours are passed through all the condensers, or only a part of them, accordingly as the operator wishes to have the alcohol more or less pure. In order that the alcohol should not evaporate in passing from the worm into the hogshead, &c., and that the stream of the liquor may be seen at the same time, a pipe is attached to the extremity of the worm, communicating with the bunghole of the hogshead.

The terminating part of this pipe is formed of glass, through which the liquid may be distinctly seen. This instrument is called the *lantern*. The alcoholic vapour that passes into the first egg in a state of ebullition, and deposits a part of its caloric there, contributes to the ebullition of the wine in this vessel, and disposes the liquor to distillation; still the wine is not carried to that degree of heat necessary for this operation till a considerable time after the distillation has commenced from the still. It is then less pure than when it was first put in; it is charged with watery vapours that have not been able to combine with it.

Two different products are then brought up to the superior part of the first egg; that is to say, the brandy that came out of the still, but disengaged from its watery parts, and the brandy produced from the liquor of the first egg. This being charged with more water than the first, weakens the first liquor; and nothing is obtained from this mixture beyond a brandy of 14° or 16°. In the passage of the liquor into the second egg, the same phenomenon takes place; but here the aqueous vapours mingle with the wine, and the alcoholic vapours rise from the second egg with a less quantity of water than those of the first, and the brandy flows at 18°. When it is the object to extract brandy only at Holland proof, or 18°, the still and two eggs are sufficient.

The cock which transfers the vapours of the second egg to the third is then closed, and that which communicates the vapours of the second egg to the highest worm, or the first worm, is then opened. The products of the still are taken till it is perceived that the liquor is diminished in strength.

The first hogshead is then removed, and replaced with another, to receive what are called *repasses*, or *feints*, in order to redistil them; and continue the operation till the still no longer yields any spirit. To know the precise moment when the distillation should be stopped, they open the first small cock on the side, which conducts to the little worm placed upon the stove, and close that which conveys the vapours from the still into the first egg. The vapours being condensed in the small worm, the liquor is received in a small glass; being thrown upon the head of the still, a piece of paper may be lighted by this hot liquor, which, if it does not burn, it is thought proper that the distillation should be stopped.

French distillers use the same process, in order to judge of the strength of the vapours disengaged from the eggs employed. When these, which proceed from the still, no longer contain any alcohol, the fire is extinguished, and they let out the residuum, which is become useless; and afterward do the same with respect to the eggs. But if, on the contrary, alcohol is still found, it is passed from the egg into the *cucurbit*, which is charged as at first; and they finish at a convenient time by adding the feints, or some wine, if it should be necessary. The eggs are then charged with the wine found in the first worm, which has already been heated in the first distillation: this is a great saving of fuel, and hastens the operation. In small distilleries, where only three eggs are used, when they would charge the eggs or the alembic with brandy or feints, they may distil three-six, by charging one or two eggs, or the alembic, with brandy or with the feints. They use a large tube, which being fixed between the still and the first egg, communicates with another, used to charge the alembic with wine; a funnel is introduced into the orifice of this tube, and by this means, and by closing the communication with all the rest, the liquor is conveyed into the vessel intended, and the cocks are also closed. The large tube here alluded to is the *corne d'abondance*, or horn of plenty. Another point is very essential to be attended to. It has been said that the tun filled with wine, in which the first worm is placed, was hermetically closed; but notwithstanding this, it receives the alcoholic vapours while very warm, and the wine is heated by them, and consequently, as well as the eggs, disengaged from the vapours.

To retain them the tun is completely covered; but in order that they may not force the cover, and thus cause the loss of the goods, the cover is made in the shape of a dome, surmounted by a small tube, which either conducts them into the worm, into the eggs, or into the still.

Observing these precautions, no loss can attend the process of distillation. With the aid of the pump the wine is conveyed from the *tampot* into the tun, and is discharged at the bottom of this vessel.

The cold wine, heavier than warm, always occupies the

lowest place, and expels the warm liquor which serves to charge the still or the egg. This construction has another advantage, as the alcoholic vapours that escape the tun can find no other issue but through the tube, which carries them into the egg.

The whole knowledge of distilling apparatus consists in *the perfect understanding* of the application of heat, of vaporization, and of condensation. For the purpose of acquainting the distiller more perfectly with his calling, all the various apparatuses and improved processes will be given, as far as thought strictly practical and useful.

It now remains to give a description of the different systems on which the most remarkable apparatuses of distillation have been constructed. These systems may be reduced to four principal and distinctive:—1. Distillation by the simple apparatus. 2. Distillation by the wine-warming condensing apparatus. 3. Distillation by steam and by rectifiers. 4. Continuous distillation. The three first will be described elsewhere in this work; the fourth will now be considered, constituting what is termed

CONTINUOUS DISTILLATION.

The continuous apparatus, which is here to be described, fig. 2, has undergone many improvements, and is now presented in its most perfect state.

This apparatus is composed—

1st. Of one still, and sometimes of two.
2d. Of a distilling column.
3d. Of a rectifier.
4th. Of a wine-warming condenser.

CONTINUOUS DISTILLATION.

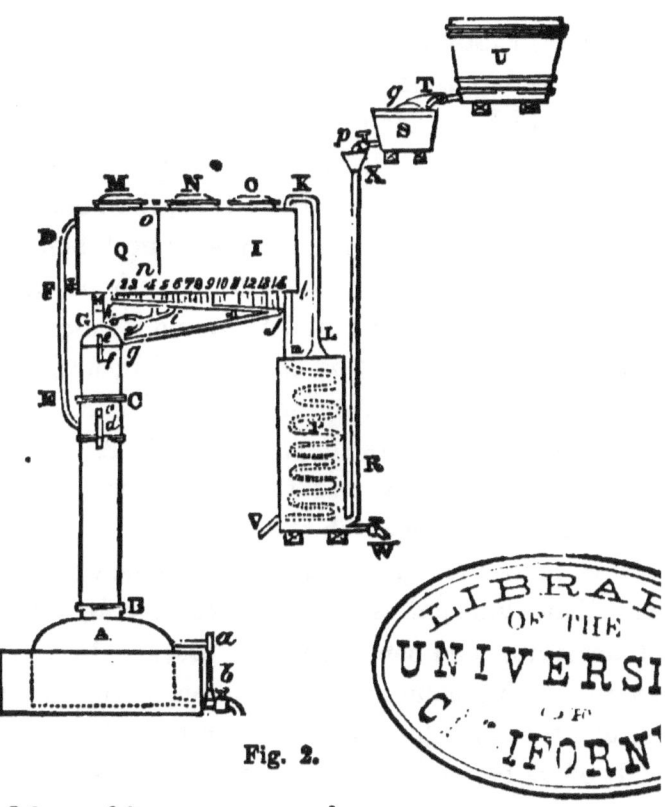

Fig. 2.

5th. Of a refrigerator, or cooler.
6th. Of a reservoir.
7th. Of a regulator, furnished with a cock, which is opened and closed by a float. Each of these parts is to be examined, and first will be considered, (fig. 2.)

I. *The Still.*—The figure shows but one still, A, although two might be used, and would be more advantageous.

II. *The Distilling Column.* In the distilling column the distillation of the wine is effected. This column, B, C, contains an ingenious mechanism, in which the wine is almost placed in immediate contact with the steam produced by the still. To this effect the steam meets with obstacles in falling, and presents itself multiplied obstacles to the ascension of the steam, which this body cannot conquer without passing through the wine; by these means the latter is divided, and so perfect a contact is established, that, in a very short time, the analysis is completed. In fact, the wine arrives almost boiling in the column, through the conduit D E, without having lost any part of its alcohol; and the more it descends toward the still A, the more it is deprived of it, until it falls in the still in a state of spent-wash. The contrary takes place with the vapours supplied the still A; on leaving it they are quite watery, and they arrive at the point C of the column in a very rich state, although this richness is always proportionate to that of the wine operated upon. The little tube $c\ d$ is a level necessary for the purpose of observing and conducting the work. It will thus be seen that this column of distillation, little elevated as it is, fills the same functions as a multiplicity of stills.

It offers better results and greater effects, with much less copper, and presents the advantage attached to the system of continuity.

III. *The Rectifier.*—This is that part of the apparatus which is marked C G; it surmounts the column, of which it is only a continuation, and contains the same mechanism. The spirituous vapours, such as they are, supplied by the column, pass through the rectifier, by the conduit H, into the wine-warming condenser, which will be immediately spoken of. There they are rendered richer in alcohol, when the spirit is required to be of great strength. This rectification is effected in the following way:—

The vapours condensed by the condenser $Q\ I$ pass through the tube $h\,j$ into the refrigerator, when they are sufficiently rich; but, if this is not the case, they may, by means of the retrograding pipes $g\,i$ and $g\,j$, be brought back, whole or only in part, to the rectifier; there they meet with obstacles in their fall similar to those opposed to the falling of wine in the column.

These low wines undergo thus an analysis similar to that which the wine undergoes in the column; that is to say, that these low wines arrive in the rectifiers much richer in alcohol than the wine that is worked, and that they leave this part of the apparatus in a state of richness about equal to that of the vinous vapours. Thus it is evident that these low wines have been deprived of their alcohol in favour of the spirituous vapours by which they have been analyzed. It is thus that, by means of the rectifier and of the retrograding pipes, the strength of the spirits may be regulated. It has already been seen with what art and ingenuity this apparatus has been constructed, and how successfully it fulfils the principles that have been established on the art of distillation.

In fact, those vapours that are the most watery are always in contact with the weakest part of the wine; and reciprocally, those that are charged with the greatest quantity of alcohol, when they are to be rendered richer, are always in contact with the richest liquid. Thus every thing concurs to deprive the wine of its alcohol without ever rendering it richer itself, and to dephlegm the vapours without ever mixing them with liquids poorer in alcohol than themselves. This advantage should be well observed, for it belongs entirely to the system of continuous distillation. The glass tube ef, the same as cd, serves to indicate the movement of the liquid in the column.

IV. *The Wine-warming Condenser.*—This apparatus, shown in Q I, like the preceding, has two distinctions :— First, to condense the vapours with which it is supplied, for the purpose of transmitting them either to the receiver or to the worm. Secondly, to appropriate to the wine intended for distillation the heat which the vapours lose by being condensed. It is evident that these functions are closely connected. This condenser is a copper cylinder, into which the wine arrives gradually through K L, to leave it through D E. It contains a vertical worm, the pipes or which all communicate, by their inferior parts, with the pipes hj and gj, through the tubes 1, 2, 3, 4, 5, 6, 7, 8, 9, 10, 11, 12, 13, 14 ; and the vapours arrive in this worm through H, on leaving the rectifier G C, which they leave entirely condensed, through the fourteen tubes, or through lm; hence they proceed either to the rectifier or to the cooler. In the execution of this wine-warming condenser conditions are to be fulfilled which are not easily surmounted; but by proper care and attention no fear need be apprehended—to such a state of perfection has the apparatus been brought.

The following are the difficulties which present themselves :—

On one side it is necessary, in this system of distillation, that the common temperature of the condenser should not exceed that of ebullition, because, if this were the case, the wine, which is much poorer in alcohol than the vapours it has to condense, could not fill this object, in virtue of the rules laid down on the capacity of alcohol, of water, and of their vapours, for heat.

On the other side, the wine, arriving through B C in the distilling column, should nearly be at the boiling point; for, without this condition, instead of being analyzed by the alcoholic vapours, it would condense part of them to acquire its maximum of heat; and this would be a real defect, occasioning a loss of time and heat; besides, the space through which it passes in the column, being calculated to operate on its analysis, admitting it enters immediately in distillation, would, in the former case, not be large enough to deprive it of all its alcohol; and a large proportion of the latter would accompany it into the boiler. Now this is what has been done to conciliate these two dissenting conditions:—The condenser has been divided into two equal parts, Q and I, by means of a diaphragm, *n o*, which, having an opening toward the bottom of the condenser, allows the wine to arrive gradually through K L, and to pass continually from I to B.

The pipes of the condensing-worm which are immersed in the wine of Q contain the most watery vapours; these, of course, abandon more heat by condensation. The wine contained in Q is warmer than that of any other part of the condenser; and, what is more, the wine which leaves it through D is always the warmest, in virtue of the laws of specific gravity. A stopcock F is fixed to the condenser, for the purpose of discharging the wine when the apparatus requires to be cleansed.

V. *The Refrigerator, or Cooler.*—The cooler P is a vertical cylinder, in copper, into which the wine is received through the conduit X R, from whence it passes into the condenser, through K L, which is fixed on the upper part of it. It contains a worm, into which the vapours are condensed, and leave through V in the liquid state. A cock W is used to discharge the worm when the working period is at an end.

VI. *The Reservoir.*—The reservoir contains the wine intended for distillation; a cock p is fixed to it; the degree of aperture of the latter is regulated by the quantity of wine with which the apparatus is to be supplied in a given time. But as this quantity may often vary, according to the unequal pressures caused by the unequal heights of the liquid contained in the reservoir, the height and pressure are consequently regulated by means of the following regulator.

VII. *The Regulator.*—U, is a small vessel into which the wine is introduced, either by means of a pump, or runs into it naturally if it can be so contrived. Its inferior part is provided with a cock, which opens or shuts according as the liquid sinks or rises in the reservoir. This result is obtained by means of a float q.

MODE OF WORKING THIS APPARATUS.

It is filled through U, which is the highest part of the apparatus. Thus the wine comes into the still, and fills it to the height required, which is indicated by the glass level; then the distilling column is charged with that portion of the w ne which is to oppose the passage of the steam. At this period the condenser and refrigerator are full; the introduction of the wine is suspended for a time, and is again continued by opening the cock *p*, to supply the apparatus with a continuous stream of wine; this is only done when the wine in the still has been entirely deprived of its alcohol, and when the wine which is in the condenser is sufficiently hot to be introduced into the column.

Then begins in reality the continuity, and all the previous work is only preparatory, although distillation has already begun. There are two very distinct parts in this apparatus; one is that in which the steam, mixed with the boiling wine, or with the low wines also boiling, undergoes, by means of this mixture, a change which is the most conformable to the object of distillation; the other is that in which the vapours are only in contact with the wine through the intermediacy of the worms in which they are condensed, and their heat is abandoned in favour of the wine intended for distillation. The first is evidently composed of the distilling column and of the rectifier; the condenser and the refrigerator constitute the second. To account for the effect of the first part, the rules laid down on the various capacities of water, of alcohol, and of their vapours for heat, must be borne in mind. Water when arrived at 212° cannot take any more heat without being transformed into steam; it occupies then a volume one thousand seven hundred times greater, and although the steam possesses the same temperature as the water by which it has been produced, that is to say, that it does not cause the thermometer to rise above 212°, yet it contains eight times more heat than water; for about two pounds of steam mixed with fourteen of cold water gives sixteen pounds at 212°. When pure, alcohol--

that is, when weighing 152°—passes into vapour at 172° temperature.

Its vapour possesses the same temperature, and contains much less heat than the vapour of water; for two pounds of alcoholic vapour, mixed with about six of cold water, will only give a mixture of alcohol and water of 172° heat. Vapour of water, which can only remain vapour at 212° of temperature, will be condensed at a temperature at which alcohol will keep its vaporous state: in water, for instance, of 172° temperature, the vapour of water will be condensed, when, at the same time, that of alcohol will pass through it without undergoing the least condensation.

If, instead of passing through water at 172°, this vapour passed through boiling wine, the water will be condensed in favour of the alcohol of the wine, which will be vaporized in relative proportions, and this in virtue of the well-recognised fact that when wine, composed of a mixture of alcohol and of water, is in a state of ebullition, alcohol only takes the temperature of 172°, which is, of course, colder than that of water. What happens in this case? The vapour of the water, in traversing the mixture, is condensed, because it meets with alcohol which has only 172°; and as the latter cannot take any more heat without passing into vapour, it is vaporized by means of the heat which the steam of water has abandoned in being condensed

Supposing the vapour which passes through wine in a state of ebullition to be itself a mixture of vapours, of water, and of alcohol, it is easily foreseen what will happen,—the portion of alcoholic vapour will pass without losing any thing in the wine, while the portion of watery vapour will be condensed, and produce a relative quantity of alcoholic vapours. Such are the phenomena which take place in the systems in which one still is distilled by the other. Such are, also, the phenomena which are observed in the distilling column and in the rectifier of the apparatus now under consideration. The nearer the vapours are to the summit of the column the richer the wine they meet, and the more they are charged with alcohol.

As, in this case, the wine operated upon, and such as it is supplied by the condenser, is the richest, and as these vapours are greatly charged with alcohol when they leave the column to enter the condenser, it must be conceived that this column has an immense advantage of other stills; and that it serves only and continuously to enrich the vapours, without ever enriching the wine; while in other apparatus it is always necessary to render the wine rich before richer vapours can be obtained. The same phenomenon takes place in the rectifier. The low wines, which run back into it, present to the vapour a liquid much richer in alcohol than that which it has met in the column; but these low wines only appropriate to themselves the water of these vapours, to which they abandon a portion of their alcohol. The spirituous vapours, on leaving the rectifier, enter, through H, into the worm of the wine-warming condenser : even in this part of the apparatus they may be more dephlegmed, and from these they pass into the worm. In this apparatus every thing is combined in such a manner as to cause all the vapours that are produced to be condensed in the wine-warming condenser; there they take the liquid state, but as they are in contact with wine which they have already rendered very warm, they cannot be cooled there. They are suitably cooled in the refrigerator, where they are brought into contact with cold wine.

The advantages offered by the apparatus now under consideration are—

First, to be able, within a given time, according to the size of the boiler, to distil a much greater quantity of wine than can be done by any other apparatus, depriving the same of all the spirit it contains.

Secondly, to be managed easily and without much labour, as there is no necessity for repeatedly charging the still; for two men may at once direct two or three of these machines without fatigue, having no other charge than that of watching and supplying the fire with fuel, which, considering the small quantity made use of, is not very laborious.

Thirdly, the whole apparatus can be had at a very moderate rate, compared with many others, and it produces more spirit than any of them.

Fourthly, it occasions a great saving in fuel.

Fifthly, being simple in its mode of construction, not much room is required; it is not liable to obstructions; and is easily repaired, supposing it, which is not the case, capable of derangement.

Sixthly, to furnish at will spirits of a superior quality.

Seventhly, not the least quantity of water is wanted for the condensation of the vapours or to cool the spirits, the matter intended for distillation being always sufficient to absorb the heat of the whole of the vapours produced

CONTINUOUS DISTILLATION. 37

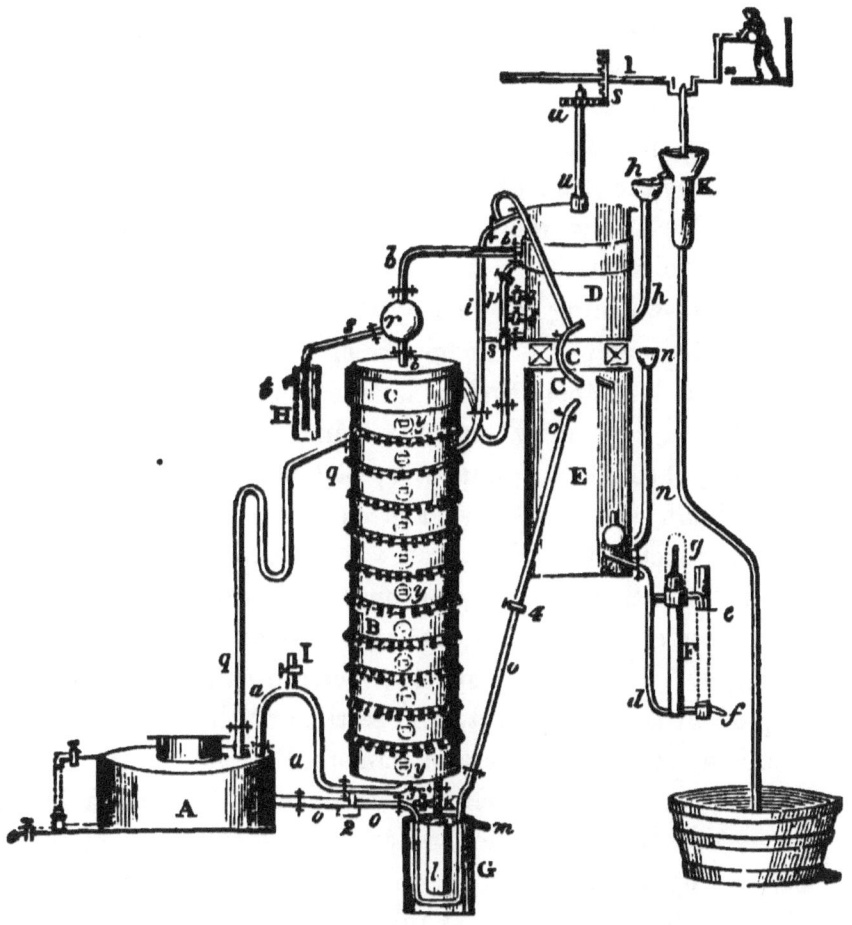

Fig. 3.

The annexed cut (fig. 3) is an improvement on this apparatus; the alterations, &c. which, it is said, render it more adapted to the distillation of wash, are fully explained in the following description of it.

A, boiler producing the steam which is to effect the distillation. B, distilling column, divided into ten pieces, each of which has from one to three screwing stoppers

y, y, y, by which means the inside may be seen, and the hand introduced in every part of the interior. c, rectifier, placed on the top of the column, in which the low wines, returning from the condenser at the distiller's leisure, are rectified by the steam of water ascending from the column; the residue of the rectifier leaves the tube $q\,q$, and runs into the boiler A.

D, wash-warming condenser; the conduits of which, intended to receive the steam, are of a peculiar form, presenting a large condensing surface. The matter to be distilled is constantly introduced into it by means of a pump k, through a funnel $h\,h$; the vapours arise from the column, are condensed, and heat this matter to 80°. The products of condensation are divided by the stopcocks 7 and 8, which may be opened to send back into the rectifier those that do not possess the requisite strength. $u\,u$ serves to stir the matter in the condenser, to prevent the heavy part from settling at the bottom.

E is a common refrigerator immersed in water.

F is a probe, into which the produce of distillation is received. In the middle branch g, covered with a glass bell, an hydrometer and thermometer are placed; $e\,f$ is a branch formed by a glass tube; the products of distillation are seen running through the small tube e; they run through f into the vessels intended to receive them.

G is a vessel into which the spent-wash falls; it comes in through k, and runs out through m. This vessel fills the functions of an hydraulic safety-valve.

H is a vessel with a plunging-tub; in case (which is not at all likely to happen) the matter in distillation should ascend into the column, it could not proceed to the condenser, for it would be stopped in the ball *r*, and through the tube *s* run out by *t*, while the vapours should take their direction to the condenser. I is a mechanism formed by a banded axis and two wheels with teeth; it is moved by one man, who causes the pump *k* to play, and turns the shaft *u u*, to the bottom of which two wings are fixed, for the purpose of continually agitating and preventing the matter from settling at the bottom of the condenser. K is a pump, which brings the matter from the jack back into the funnel *h h* of the condenser.

MODE OF WORKING THIS APPARATUS.

The still A is filled with water, (the first time the column and condenser are filled also with water;) the water in the still is brought to ebullition; the steam passes through *a a a* into the inferior part of the column, ascends from case to case, passes through the rectifier into the condenser, where it abandons its caloric in favour of the water contained in the latter. When this condensed water arrives at the probe F, the pump K works without interruption.

The matter proceeding out of the pump having sent the water out with which the condenser has been filled, arrives in the column through *p p*, where it is met by the steam, which causes it to boil; it descends from case to case in a constant state of ebullition, and, arrived into the last case, it runs into G, and leaves through *m*. By opening cocks 7 and 8 of the condenser, the lowest products of distillation are sent back into the rectifier; there they are dephlegmed, and return at a very high strength, which does not vary during the whole time of distillation. As the water which passes in the state of steam out of the still A is to be replaced, and as it is indispensable that this water should be very hot, it is drawn from the worm-tub E through *o o o*; before it enters the still it passes through G, where the matter is boiling hot. Cock No. 4 regulates the quantity of water which is to be introduced.

When all the matter has been pumped out, the process is continued for about a quarter of an hour, for the purpose of exhausting the matter left in the column. The fire is then drawn off; cock No. 1, fixed on the tube *a a a*, is opened, as also cock No. 3, which is fixed to the tube K on the lower part of the column. The next day the operations are recommenced in the same way. The column and the condensers are left filled with matter.

There is another system of continuous distillation which will here be described; it possesses the advantage that it can be applied (the distilling column) to the neck of the still. See fig. 4, on the opposite page.

Description of the Apparatus.—A, double still, having a copper partition in the middle, which divides it from top to bottom; there is a hole at the lower part of the partition, so as to establish a communication between the two stills.

B, first distilling column.

C, second distilling column.

Each of these columns contain an evaporator in the form of a double vice of Archimedes.

D is a common capital or head belonging to the first column.

E, $\frac{2}{6}$ capital or condenser, for the purpose of making strong spirits.

CONTINUOUS DISTILLATION. 41

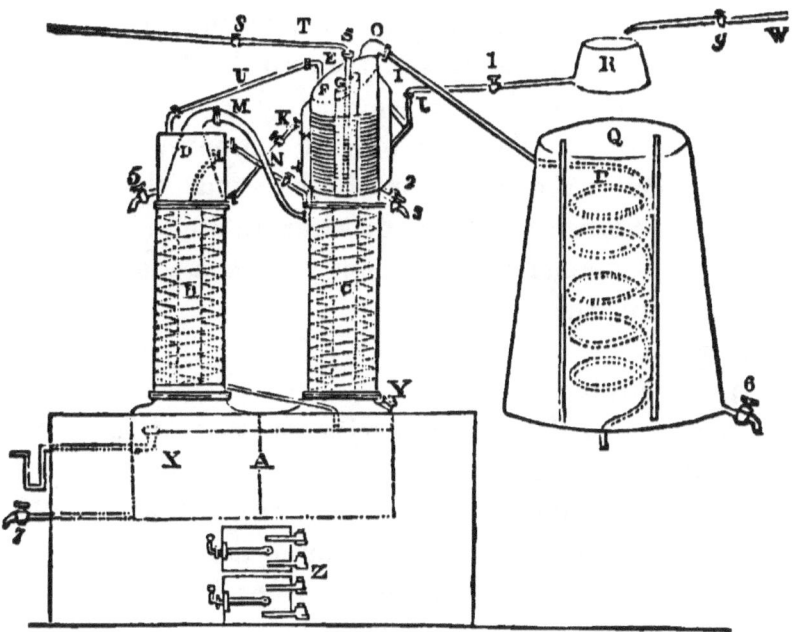

Fig. 4.

F, column of the head, containing an evaporator, constructed in the form of Archimedes's vice.

G, space left in the middle of the head to receive the water necessary for the condensation of the low wines.

H, envelope of the column which receives the cold wine through the funnel J.

I, tube, having a regulating cock; it takes the wine from the tub R into the funnel J.

J, funnel which takes the wine in the envelope H by means of a double-branched tube.

K, tube which takes the wine from the envelope of the head into the lower part of the common head.

L, tube which introduces the wine of the envelope of the common head on the evaporator of the first column.

M, tube which takes the spirituous vapours of the first column into the upper part of the second, which sends it into the pipes contained in the head.

N, tube which takes the wine of the envelope of the first column on the evaporator of the second; this tube is only used where spirits of a low strength are to be made.

O, tube by which the spirituous vapours are taken from the head into the worm.

P, worm.

Q, worm-tub.

R, supplying the apparatus with the wine; it is to be alimented by a larger tub.

S, funnel, followed by a tube which takes the cold water into the middle of the column of the head.

T, tube which takes the water into the funnel s.

U, air-tube.

V receives the spent-wash at the bottom of the first column, from which it proceeds into the second still through a pipe.

X, funnel communicating with the outside of the still by means of a tube; it is used for the purpose of evacuating the spent-wash, which, when it is above the level of the funnel, runs out of the still.

Y, man-hole.

Z, doors of the furnace and ash-hole.

W, tube which aliments the tub R.

No. 1, regulating cock, supplying the apparatus with wine.

2, cock to discharge the envelope of the head.

3, cock to discharge the middle of the column of the head.

4, cock which takes the wine of the envelope of the common head on the evaporator of the second column.

5, cock to discharge the envelope of the common head.

6, cock to discharge the worm-tub.

7, cock to discharge the still.

8, cock to bring the water.

9, cock to aliment the tub R.

MODE OF WORKING THE APPARATUS.

The still is filled with water, which is brought to ebullition. When the distilled water runs out of the worm the operation should commence. Cock No. 1 is about half opened, so as only to give passage to half of the wine it is capable of furnishing. The interior G of the head is filled with water by cock No. 8. The wine runs through the funnel J into the middle of the envelope H, which is filled as high as the tube K; it enters then through this tube into the lower part of the envelope, and, when the latter is full, the wine runs from this part through the tube L on the evaporator. The wine runs on the evaporator of the first column on which it is distilled; and the spent-wash runs out at the bottom of the column. The spirituous vapours rise into the head D, and proceed through M into the upper part of the second column; they then rise into the pipes of the head, in which they may be more or less condensed; the spirituous vapors alone can traverse it, and the phlegms return back into the first column and are again distilled before coming into the still. In this operation the wine is distilled on the evaporator of the first column, the phlegms on the evaporator of the second, and the spent-wash falls into the still, which is the most distant from the evacuating funnel x, where it remains long enough to be entirely deprived of its alcohol.

The various strengths are obtained by the degree of cold given to the head by means of the water introduced by cock No. 8. In the apparatus as previously described, the wine is in immediate contact with the steam in the column, but in this last the contact is effected through the medium of the coppers of the evaporator.

APPARATUS USED PRINCIPALLY IN AMERICAN AND ENGLISH DISTILLERIES.

As regards the vessels mostly used in this country and England, when the condensed vapours are obtained in the liquid form, the shape and situation of them are very different. The vapour should be kept completely in its elastic form, to a certain height. The neck of the vessel should then turn by a sharp curve on an elbow, so that the substance, after condensation in the liquid form, may, by its gravity, descend as quickly as possible. The height of the elbow above the point where the heat is applied should be only sufficient to guard against the mass below getting over the neck by boiling. When the neck of the lower vessel is liable to be long, it should be defended either by being polished or clothed, to prevent the escape of heat, in order to allow the vapour to be carried over into the descending part before it condenses.

The vessel from which the vapour rises, when of a

large size, and used for distilling simple fluids, is called a still. Those for experiments in a small way, and also for distilling acids, ammonia, ether, &c., are called *retorts;* the vessel that receives the distilled matter being called a *receiver.*

When the worm-tub is employed, the still requires such a shape that the greatest possible surface may be exposed to the fire. Its shape is then that of a frustum of a cone. The neck should be of such width as to convey the vapour away as fast as formed. The height of the neck is regulated by the nature of the substance operated upon: if it is mucilaginous, the neck should be longer, to prevent its boiling over; and the exterior of the lower part should be polished to prevent the escape of heat; the descending part painted black, and its end inserted into the end of the worm. The worm-tub is a wooden vessel, about six or eight times the capacity of the still; the length to the diameter should be about ten to seven. The worm consists of a spiral tube, which enters on one side of the tub at the top; it then passes spirally, in six or eight convolutions, to the bottom, where it comes out of the side in order to discharge the liquid arising from the vapour condensed within it by the agency of the cold water with which the tub is filled.

The water is constantly changing, by the warm water running away from the top, while a supply of fresh water comes in at the bottom. The section of the tube being a circle, its capacity should not be less than one-fifth of that of the still; the diameter of the ends of the tube being about three to one. The object is not merely to effect a condensation, but to cool the liquid: the means of performing this will not always depend upon the relative size of the vessel, but must be governed by the quantity of vapour supplied in a given time, and the supply of cold water. The vessel that encloses the substances to be distilled is called an *alembic*.

It is very dangerous to use one that is not tinned, as the liquor would assume a deadly quality. An alembic is composed of two or three parts, according to which the distillation is carried on by fire or by the *Balneum Mariæ:* in the first, the bottom of the alembic is in close contact with the fire; but in the latter, its lower part is placed in another vessel larger than itself, which, being filled with water, acts as a medium between that and the fire. The alembic, properly speaking, is composed of two parts, the cucurbit and the head; but though the form of the latter may vary according to the systems of operations adopted, its use is always the same, namely, to contain the matter intended for distillation. When the cucurbit is large and spacious, as it must be in great distilleries, then it is necessary to fix it in the masonry of the copper.

In this case it is difficult to clear it of the phlegm, or the residue of the distillation, even with the aid of a siphon, according to the practice of the ancients. However, this is now best remedied by a cock on one side of the vessel, near the bottom, that must be set running when the alembic is discharged, or when it is necessary to clean it. The size of the cucurbit varies in different countries. Many distillers, to augment their products, or to ameliorate the quality of the liquor, add a cooler to the head of the still. However, the observations continually made in large distilleries have sufficiently proved that coolers, if not pernicious in the distillation of brandies, were useless. Consequently the use of them was dropped, as was also the pipe or tube called the *blackamoor's head*. But whether the still-head be conical or otherwise, its uses are always the same, viz. to receive the vapours caused by the ebullition of the liquid, and to transmit them through the different tubes that surround the still-head and form a part of it. These tubes present the figure of a truncated cone, the smallest diameter of which is the most distant from the head. Every vessel composed of copper in a distillery should be well tinned, and continually examined; otherwise a deterioration will occur. The acid of wine as well as that of ardent spirits, corroding the copper, will form verdigris, which will be mixed and distilled with the liquor.

In describing the stills of this country and Great Britain, it is necessary to observe that all distillatory vessels are either alembics or retorts. The former consists of an inferior vessel, called the *cucurbit*, designed to contain the matter to be examined, and having the upper part fixed to it called the *capital*, or *head*. In this last the vapours are condensed by the contact of the surrounding air; or, in other cases, by the assistance of cold water enclosing the head, in a vessel called the *refrigeratory*, or *cooler*. From the lower part of the capital, or still-head, a tube proceeds, called the *nose, nozel, beak*, or *spout*, through which the vapours, after condensation, are made to flow into a vessel called the *receiver*, which has usually been spherical.

Receivers have had several names, according to their

figure, being called matrasses, balloons, &c. There have been various modes of applying heat in distillation, depending upon the nature of the apparatus employed as well as upon the substance to be distilled. The common still being formed of metal, is immediately exposed to the naked fire, since from its tenacity, and its property of conducting heat with facility, it is not liable to crack, which is not the case with glass or earthenware. The still is heated in various ways, the most common of which has been by the *sand-bath*, a vessel of iron filled with fine dry sand.

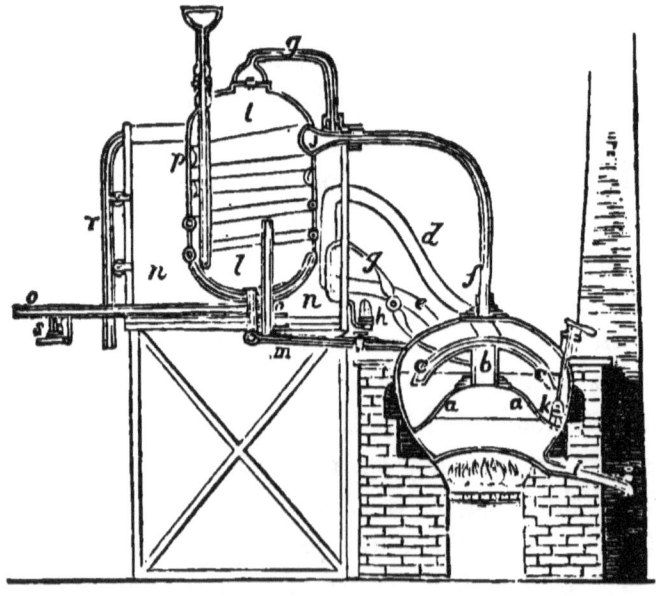

Fig. 5.

The annexed figure (5) is a specimen of a still which has stood the test for a *number* of years, and is one worthy of the attention of the reader.

The still here represented is made of metal, as usual, but having one internal division $a\ a$; this forms the still into two compartments; these are both charged with wash, or other liquid intended for distillation. The still being heated by the furnace below, the vapour from the lower compartment will be driven through the tube b, and descend by the bent pipes $c\ c$ into the wash of the upper compartment, or from the tube b, without the bent pipe; the vapour may be dispersed above the surface of the liquor within the still, by the intervention of a plate placed over the tube b, by a pipe d, up to the vessel of water, and descend again by the pipe e into the upper compartment of the still; by which means a partial condensation will have been effected of the grosser vapours which have arisen from the lower compartments, and the higher or uncondensed vapour will pass off through the perpendicular pipe f to the condensing apparatus. A small pipe g, with a stopcock, is inserted into the tube b, and carried through the vessel of water, by which a small quantity of the vapour from the lower part of the still may be admitted into the glass vessel h, for the purpose of ascertaining the quality of the vapour.

When the spirit is out of the lower compartment of the still, and the upper compartment reduced to the gravity intended, the spent liquor below is to be drawn off through the cock i; after which the valve k may be opened, to admit the liquor from the upper to the lower part of the still; and the succeeding charge is drawn from the cistern l, through the pipe m, to the upper part of the still. The condensing apparatus consists of two cisterns, placed one within the other; the inner one l

should be made of copper, and is filled with wash or other liquid intended for subsequent distillation, by means of the pipe and funnel.

This vessel should at least contain two charges for the upper compartment of the still. The outer vessel *n n* may be made of wood, and must be charged with water circumscribing the inner vessel. When the vessel is attached to other stills, the outer vessel *n n* may be dispensed with, and the pipe *o o* attached, which must communicate with another condenser. The vapour passing from the still through the pipe *f*, as above described, proceeds to the spiral condensing pipe *p*, which passes several times round the vessel *l*, and the vapour being cooled and condensed in its progress, finally collects in the form of spirits, and is drawn off through the pipe *o*; *q* is a pipe inserted into the head of the vessel *l*, with a light valve opening upward, for admitting any vapour into the condenser *p*, which might arise from *l*; but this at the same time prevents the passage of any vapour from the still through the pipe *f*. A waste-pipe *r* is attached to the cistern *n*, to carry off the surplus water. The parts of this apparatus claimed as an improvement by a gentleman in London are a vessel to contain wash or other liquid in the progress of distilling, surrounded by a cavity for condensing the water from the still. The apparatus shown at *h* and *s* are glass vessels containing a number of graduated bubbles of known gravity, which being put into a glass vessel, into which a portion of the spirit passes from the still according to the number of bubbles afloat, denote the levity of the spirit by their colour or shape.

With the assistance of a thermometer, the strength is

ascertained. Here it is not necessary to describe the well-known graduated bubble. Another improvement made on this is to cause the liquid operated upon in the process of distillation to flow gradually over the heated surface of the boiler while it continues to give out its spirituous evaporation. The quantity of liquid allowed to be acted upon, or to pass through the still in a given time, and also its velocity, is regulated by the circuitous route in which it proceeds; and by that means the complete operation of the fire upon the whole fluid is insured, without impeding or clogging the spirituous evaporation by aqueous or empyreumatic vapours.

By this construction of a still, a continued and uninterrupted distillation, boiling, or evaporation is carried on as long as the supply of liquid is furnished and the fire kept up.

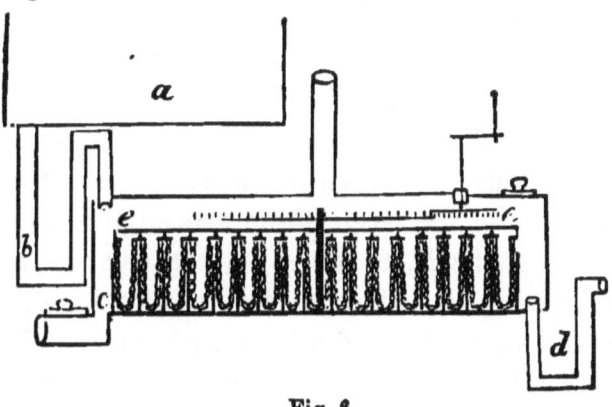

Fig. 6.

In fig. 6 is a view, in profile, of the section of a still or boiler made on the improved principle, of copper or any other suitable material; and fig. 7, on the following page, is a plan of the same.

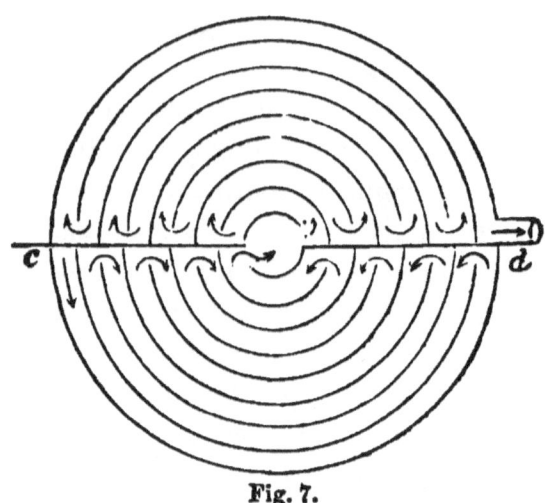

Fig. 7.

The bottom of this boiler is divided by concentric partitions, which stand up (as shown in fig. 6) sufficiently high to prevent the liquor from boiling over. These partitions have openings from one another at opposite sides, so as to make the course a sort of labyrinth. *a* is a reservoir of liquor prepared for the operation; *b* is a pipe or tube descending from the reservoir, conducting the liquor to that part of the boiler marked *c*, which is the commencement of the race. From hence the liquor flows through the channels, as shown by the arrows, progressively traversing the whole surface of the bottom; so that the full effect of the fire is exerted upon small portions of the liquid, which causes the evaporation to proceed with great rapidity.

The residue of the liquor then passes off by the discharge-pipe *d*, contrived to slide, for the purpose of regulating the quantity and depth of the fluid intended for the still; and this pipe should be in such proportion to the admission-pipe as to cause the perfect distillation of the liquor in its passage to the regulating tube.

The spirit which rises in the head of this improved still will be found much stronger and purer than that obtained from stills of the ordinary construction, where the spirituous vapour is much mixed with aqueous matter and other impurities. The channels may be extended to any length required, over a bottom of any dimensions, by contracting their breadth. Stills upon this principle may be made of all sizes and shapes, round, square, or otherwise; and the partitions may be placed in concentric or eccentric circles, with openings on their sides at such distances as shall cause the liquor to flow over the most extended surface of bottom; or the still may be square, with angular partitions ranged as a labyrinth, or in any other manner, so as to cause the run of the liquor to be greatly extended over the surface of the boiler.

The bottom of these stills may be either flat, concave, convex, conical, or of any other form; and the entrance of the liquor into the still, and also its discharging aperture, may be at the side, in the middle, or elsewhere, as circumstances may dictate. Boilers or evaporators may be made on this plan, either with or without heads, and their capacity of working may in all cases be increased by placing layers of pipes, connected thereto, within the flues, between the still and chimney; which pipes may be bent or coiled in a serpentine direction or in any other position, and the liquor to be operated upon made to pass through them previous to its entering the still: thus the operation may be advanced to any required state of forwardness. Stills of the above description, particularly if made square, may be divided internally into several, each having its separate head and condenser; by which arrangement the spirit condensed from the first may fall into the second, to be again operated upon, and so on to a third, whereby a rectification may be carried on to any degree at one operation and by one fire. In the still shown at fig. 6 a set of chains are seen suspended from the bar *e e*, supported by a central shaft, that may be put in motion by a toothed wheel and pinion, actuated by a crank or winch.

These chains hang in loops, and fall into the spaces between the partitions, to sweep the bottom of the still as the shaft revolves; and thus they prevent the material acted upon from burning, when of a thick, glutinous nature, as turpentine, syrups, &c. Ledges may be placed between each circle, on the principle more particularly explained at fig. 8, which is square, oblong, or round, its bottom intersected with portable ledges, fastened at the ends and bottom, if square or oblong, and only to the bottom, if round; except that under each alternate ledge a space is left, of any width required, between it and the bottom, so that the liquid, in entering at the end or centre, passes over the one and under the other ledge, until it arrives at the point of discharge.

Thus the whole mass, whatever depth it may be in the still, is submitted to the full effect of the fire in a layer of the thickness of the space between the ledge and the bottom.

Fig. 9 differs essentially from all the others in this, that the bottom is doubled up and down in plaits, and represents a surface commensurate with the length, depth, and

CONTINUOUS DISTILLATION. 55

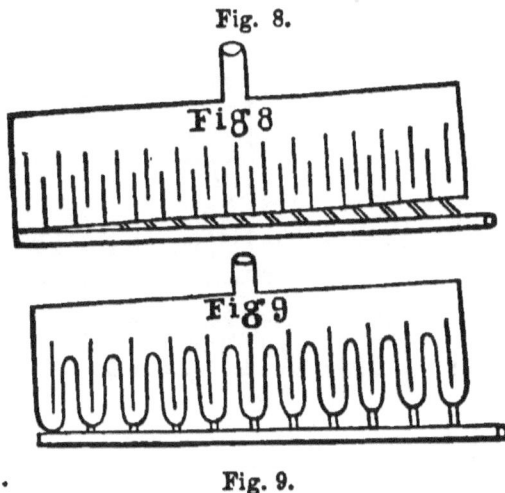

Fig. 8.

Fig. 9.

number of the plaits, between each of which (as in fig. 8) a ledge is run from side to side of the still, and fastened to both, leaving a passage the whole length underneath, between its lower edge and the bottom of the groove formed by the plait, by which the liquid in its whole course is reduced to a stratum of any thickness required along a surface of immense extent, occupying comparatively but a small space, and exposed to all the heat of the fire. In the foregoing descriptions the stills have been considered as in immediate contact with the fire; but it is proposed to work them by steam, which may be applied either externally or internally, or both, as shown in fig. 10. In this figure, *a* represents a steam-boiler, furnished with safety-valves, and supplied with water in the usual way: this boiler is surmounted by three stills upon the foregoing principles. The bottoms are perforated at certain distances throughout their whole extent, and into each of these perforations a tube is inserted, branching into ramifications of smaller

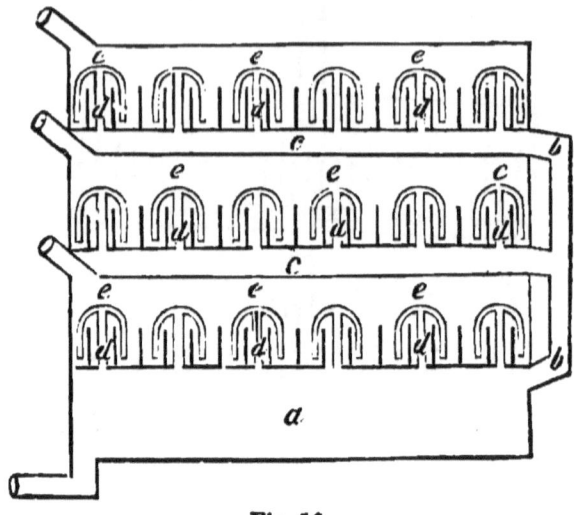

Fig. 10.

tubes the extremities of which are bent down into the liquid flowing through the still.

The steam from the boiler passes up the tube $b\,b$ into the hollow vessels c, c, and thence through the tubes d, d, d into the smaller curved pipes e, e, e; at the extremities of which it pervades the liquid in its progress. If it should be deemed more advantageous to transmit the caloric from the steam through the metal, without allowing the steam to pass into the liquid, it may be done by closing the extremities of the curved pipes e, e, e, and placing them in a horizontal position, with a small inclination, in order to allow the condensed steam to pass into the boiler.

Here the spirit arising in one still might pass into another, and be again operated upon; and distillings of every degree and of various substances may be carried on in one continued operation at the same time and by one moderate fire, which, upon this principle, will suffice for the largest establishment known. Another improvement, advantageous to the art of distillation, is an apparatus for preventing the loss of alcohol or spirit during the vinous fermentation. (See fig. 11.)

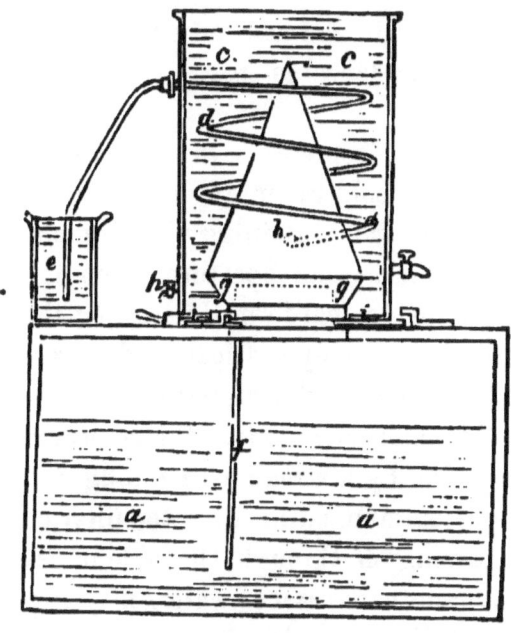

Fig. 11.

This apparatus consists of a vessel or head, constructed so as to be capable of attachment to and communication with the back or vat in which the process of fermentation is carrying on, in the production of wine, brandy, beer, &c. The back or vat is to be closed on all sides, air-tight, except an opening in the top, which communicates with the head above mentioned. This head is to be surrounded by a vessel of cold water, in order that the alcoholic vapours evolved during the process may, on rising up into the head, become condensed, and then trickle down the inside of the vessel and descend into the vat.

By the application of this apparatus a certain proportion of the alcohol, which has been hitherto suffered to escape with the non-condensable gases in the form of steam, will be condensed and returned into the liquor; while the non-condensable parts will be carried off through a pipe.

The cut (fig. 11) represents this improved apparatus, the vat and the cold-water reservoir being shown in section. a is the vat containing the fermenting liquor, in the top of which is an aperture communicating with the interior of the conical-formed vessel b; the lower part of this vessel is made cylindrical, and passes through a circular plate, on which the supporters rest. c is the reservoir of cold water surrounding the conical vessel, which may be supplied by a stream of running water. d is a worm or pipe communicating with the interior of the vessel b, and, passing off through the side of the reservoir, descends into another vessel of water, e. f is a small pipe which proceeds from the lower part of the vessel b, and descends through the fermenting liquor nearly to the bottom of the vat. The gas and alcohol rising from the liquor in the vat into the conical head b, and coming in contact with the cold sides of the vessel, produces a condensation of the alcohol, which runs down the side of the cone into the circular channel g at its base, from whence the alcohol passes by the pipe f into the vat below; while the non-condensable gases pass out through the worm-pipe d, and finally escape by bubbling up through the water into the vessel e.

If any portion of the alcohol should pass up the worm-pipe, it will become condensed in its progress, and by the position of the worm will be enabled to run back again, and pass into the vat. A small cock *h* is placed at the bottom of the cone, for trying the strength of the condensed alcohol. This apparatus may be removed from its present situation to another fermenting vat by drawing off the water and disengaging the head *b* from its place. The plate is furnished with circular wedges round the circumference, as seen at *i, i*, made to act beneath hooks; the apparatus, being turned round by the handles in a horizontal direction, becomes fixed in its place, having between the plate and the head a ring of thick leather, to prevent the gas from escaping. The sole object and novelty proposed in this improvement is to prevent the loss of alcohol in the usual process of fermenting liquors in open vats, and to return the condensed alcohol into the liquor again.

INSTRUMENT TO PREVENT INEQUALITY OF HEAT IN DISTILLATION.

DISTILLATION consists principally of two operations, viz. the conversion of the matter into vapours by heat, and the condensation of the same vapours by its opposite. Therefore, that this twofold operation should be effected with promptitude, and at the least expense for combustibles, it is necessary that a perfect equilibrium should be established between the heat in evaporation and the condensing cold in resisting the latter, by means of a given quantity of water of fixed temperature passing through the cooler in a given time. It is necessary that the fire should be regulated in such a manner that the quantity of vapour produced should be neither greater nor less than that which at the same time may be condensed by the application of cold.

The failure of attention to this circumstance (particularly in the distillation of spirituous liquors) may produce the following inconveniences:—First, if the fire is too violent, a great quantity of the condensed vapours will pass from the worm into the external air, and occasion the loss of the matter distilled, and also of the fuel. Secondly, if the fire or heat is diminished too soon, the condensation will produce a vacuum in the worm and in the alembic, which, not being proportionably filled by the fresh vapour, will admit the entrance of the external air, and impede both operations; and, lastly, will carry with it a part of the vapours, and occasion loss of time, and also of the matter distilled. To remedy these defects, and at the same time provide simple and effectual means for indicating the exact state of the heat every instant, an instrument has been invented which may be adapted to any distillery or apparatus, and is, in reality, nothing more than an application of known and practical principles. This ingenious machine possesses another singular advantage; namely, that in intercepting the communication of the atmospheric air, the products of the distillation are more abundant and perfect; for, in proportion as the vapours condense, a vacuum is formed in that part of the apparatus into which the vapours are drawn, in the same manner as by the pump: they are also less compressed

PREVENTION OF INEQUALITY OF HEAT. 61

in every part of the apparatus which they may fill; the condensation is more rapid, and the products, upon the whole, greatly superior.

To render this instrument still more useful, its lower part should be completely immersed in a vessel filled with cold water up to the ball; the liquor, though ever so little impregnated in traversing this cold fluid, will complete the deposition of its caloric.

This precaution will also prevent the losses that frequently result from the negligence of the workman.

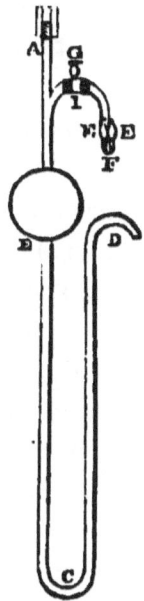

Fig. 12.

Explanation of the Cut.—A B C D, a tube of copper or glass, in several pieces, bent over, with a ball H eight inches in diameter. The upper end of the tube A may be attached to the worm by means of a vice. The length of B C, C D is four feet, and the capacity of the ball H is something more than that of the tube B C D. The distillation having commenced, the vapours condensed will pass through A and the ball H into the tube B C D. But it will only be when the two arms are filled that the liquor will go out through D to enter the vessel intended to receive it. These two arms will then remain filled during the whole process of the distillation; and in this consists the remedy of the inconveniences the instrument is intended to remove.

It is easy to see, that if the fire becomes too brisk, the uncondensed vapour will not be able to discharge itself, by opening a passage to the external air, before having driven out all the liquor contained in the tube B C, and overcome the pressure of a column the height of which is equal to C D. In the second place, the external air cannot enter to occupy the void occasioned by the slowness of the fire, but only by expelling that from D C, and surmounting a pressure of the same height. Still, this column being four feet in height, allows a sufficient latitude and time for the workmen to regulate the fires. If the tube B C D was of glass, it would only be necessary to observe the level of the liquor in the two arms.

Its being lowered in B C would indicate the necessity of diminishing the fire; and in C D it would be necessary to increase it. But as the operation in the tubes of this length is rather precarious, it would be best to attach to E a little glass regulator E F E, of which the two arms E, F, each being three inches long, contains mercury; this, in rising alternately in one or the other, would be an exact indication of the degree of the heat, and also of the vapours.

This regulator might be enclosed, so as to prevent accidents. Between this and the worm is the stopcock G, which, in the beginning of the operation, communicates with the external air in the same manner as the cock of an air-pump; but after the fire has been forcibly driven, the vapours may be seen issuing out of I; then turning this cock, the communication between the worm and the external air is closed, and the other between the same worm and the regulator must be opened, and the actions of both will commence. The ball H prevents the liquor driven by the external air from rising in E F E and in the alembic. It is scarcely necessary to add, that the head, whatever its form may be, should be well luted, in order to prevent the entrance of the external air.

OF THE PROCESS OF MALTING, &c.

For a long time, corn has supplied the trade with a kind of spirits, commonly called "spirits of corn." Among the various kinds of corn used for the purpose of distillation, rye ranks the highest. Oats, Indian corn, and wheat are also used with success; barley is almost always mixed, in a proportion which varies, with those vegetables.

The best, and we may say the only, way of ascertaining the venal worth of corn is that of its specific gravity; so that, all things being equal, that which under an equal measure weighs the most must be preferred for distillation, as well as for every other use; and the price varies, particularly, according to this quality. Its other qualities are by no means a matter of indifference; such as their perfect conservation, because those that are heated render much less spirit, their fermentation not being so good. As to the defects inherent to corn which agricultural chances have occasioned to germinate before the harvest, these are recognised by the appearance and weight of the corn; it weighs much less than that which has not undergone this change. Wheat is not much used for distillation, because, destined more particularly for human food, its value is generally greater than that of other corn, and because its produce in spirit is not proportionate to that value.

Oats, for a like reason, are seldom used for distillation, and they are useful as food for horses. Rye is the most convenient, because its produce in spirit is considerable, and also because it leaves a proper margin for the distiller. Besides, being little fit for baking, it would find comparatively little use without distillation.

There are many methods for predisposing corn to fermentation, but there exists three operations common to all; these three operations are practised in all distilleries. The first is that of grinding; the second is that known under the name of steeping; and the third that of mashing. They are of such importance in the distiller's art, that it will not be considered out of place to describe them separately, and to indicate at the same time their object and utility.

Every species of corn destined for distillation should not be ground into fine flour, but only broken. This is a practice of which experience has proved the utility; not that a greater division of the vegetable would be an obstacle to fermentation, when the following preparations are made use of, but these preparations would then be of a more difficult workmanship, and the expenses of grinding would be much heavier. These inconveniences may be avoided by only reducing corn into coarse flour. This result is obtained by having the mill-stones at a proper distance one from the other. It is customary in distilleries to use the corn as needed. This, in fact, keeps better in its natural state; it is less liable to be heated; and by these means requires less precaution to be taken for its preservation. Distillers are advised to follow this method, if they wish not to be exposed to great decrease in spirits: corn heated, either in nature or when reduced to flour, loses its fermentable properties. For the most perfect intelligence of this operation, let it be supposed that the quantity of matter to be fermented is equal to 200 pounds. The corn, being selected and ground into coarse flour, is deposited in a tub capable of holding two-thirds more than this amount, and filled so as to keep a vacuum necessary for the scum produced by the fermentation.

Then proceed to steeping. It is effected by pouring on the flour 200 pounds of water, at 120° or 130°, according to the season of the year; the water should be hotter in winter than in summer. The best mode of working is to pour on the flour a mixture of hot and cold water, such as to form, after ten minutes' brewing, a mixture at 95° or 100°, which will be easily obtained in all seasons; to effect this the use of a thermometer should not be neglected, as it is an infallible guide to conduct this operation regularly

In this state, when there are no lumps that have escaped the penetration of the water, and when the mixture has been agitated for ten minutes, the tub is left to subside for half an hour; one-quarter of an hour is even sufficient.

The object of this operation is, as its name indicates, to steep and soften the grain, by making it absorb water; and the temperature of 120° or 130° contributes to render water more penetrating—consequently, it has been recognised the most proper for steeping. If the temperature was lower, its effect would be much slower, and after a quarter or half an hour's rest the corn might happen not to be sufficiently steeped. If it was higher, on the contrary, the corn would be apt to be baked, and the operation might fail; such would be the effect of a temperature of 180°.

At this heat the fecula enveloped with gluten is baked to the surface of each fragment of corn, and forms a solid envelope, which presents an obstacle to the easy penetration of the water into the interior parts of each of the fragments; and this penetration should indispensably take place before the commencement of the following operation: a real harm results from not proceeding in such a way as to effect it. It is likewise essential, in the beginning of this operation, not to pour at once into the tub all the water necessary, and the operation will always be well conducted when the water arrives gradually, during which time the flour should be well stirred. These rules, which have just been established for steeping, are general, and admit of no exceptions, whatever be the nature and state of the grain made use of.

The steeping of the flour being finished, the next operation is that of mashing. This consists in well brewing the grain which has been steeped, while a quantity of boiling water arrives gradually into the tub, till the mixture has acquired 175° or 180°; the agitation should last five minutes, at least. At this period the tub is covered, and left to subside for a space of time varying from two to four hours. A principle may be laid down, that the longer the mixture is left to itself, the more complete will the operation be; that is to say, that four hours' standing is always better than two.

Nevertheless, it would sometimes be more injurious than beneficial to extend this space of time; such would, for instance, be the case if the mixture descended below 120°. This process will always be well executed if conducted in such a manner as not to allow the temperature of the mass, during a mashing of three or four hours, to sink below 120° or 125°. To this effect the tub should be carefully covered after the brewing is over. But it may easily be perceived that, whatever precautions be taken to avoid the loss of heat, it will always be considerable in the space of a few hours, even if there was none sustained but that occasioned by the side of the tub. The smaller the mass operated upon, the greater this loss will be, and *vice versâ*. It will always be greater in winter than in the summer; so that the talent of the distiller for mashing consists in knowing how to use the thermometer; for instance, he will give a little more heat to a small tub than to a large one—say, 145° to a tub of medium size, and 140° to tubs of great dimensions. He will also increase the heat a little in the winter, and lessen it in the summer; and in all cases the lob must be put to fermentation as soon as its temperature is fallen to 110°. Should he wait longer, he is exposed to have the whole mass spoiled by the acetous fermentation, which is easily developed at that temperature. During the subsiding of the maceration a phenomenon takes place which has for its object to saccharify the fecula of grain, and to predispose it thus to fermentation, which it could not undergo without it.

In fact, if corn, taken in a raw state, was simply diluted with water at 100°, to form a mixture bearing 77° of heat, the most proper temperature for fermentation, the latter would never be developed; or, at least, it would only declare itself after many days, and that with very little intensity. It is not the case when the mashing has been well conducted; and the more favourable the conditions under which it has been executed, the better the fermentation will proceed. Mashing may then be considered a real saccharification; and if we remark the analogy between the saccharification and an experiment by which starch has been converted into sugar, by means of water, gluten, and a temperature of 145° kept up for twelve hours, the saccharification of the fecula of corn during the mashing will easily be conceived.

In fact, all grain contains gluten, with which the starch is in immediate contact. Water is added during the operation, and the mixture is exposed to a temperature of 145°. This is the most proper temperature for mashing; by this heat starch is converted into sugar in the shortest space of time. Not that this effect could not be obtained at a lower temperature,—100°, for instance,—

PROCESS OF MALTING.

but then the mass would be in danger of turning acid, and if this temperature was to be maintained for some hours, the evil resulting from it would be irreparable.

If the temperature exceeded 145°, there would be no inconvenience to bring it to 155°, and even to 165°; but at 180° the danger begins to show itself. Above 180° there is great danger of doing harm to the fermentation; and if the heat approached 200°, there would be no fermentation produced at all. It appears that the gluten, which in this operation is the vehicle of saccharification, only posseses that property when it has not been exposed to too high a temperature; heat seconds its action very much, and renders it more intense, but the maximum of this heat is from 145° to 165°. The proportion of water acts also a remarkable part in the maceration, and the greater the bulk made use of, the more prompt and complete the saccharification will be, all other necessary conditions being fulfilled; for instance, generally, to work 400 pounds of flour, about 12 gallons of water are used in steeping, and 30 are added in mashing; this will be complete in four hours.

But if the dose of water was doubled, this mashing might be as complete as the former in the course of from two to two hours and a half; such is the influence of water on the saccharification of starch.

These phenomena will hereafter be demonstrated by other examples; when speaking of the maceration of potatoes, this will be particularly considered. When the mashing is over—that is, after two or four hours, according to the quantity of water made use of, a temperature bordering upon 145° having been maintained—then the liquid is put to fermentation. This is done by adding to it a new quantity of water, so as to have the mixture well diluted.

Previously, the operations common to the various methods made use of to predispose corn to fermentation have been signalized, and which always take immediate precedence over the latter; but it often happens that corn intended for distillation is submitted to a previous preparation, known by the name of malting. It is scarcely ever the case that an individual uses raw (that is, unmalted) corn for the purpose of distillation. The French and English distillers always mix their raw grain with a certain portion of malt, and a great many German distillers work entirely with malted grain. It will thus be essential minutely to indicate the best processes practised in malting. This operation is composed of several others, which will be successively described.

In steeping, the corn is thrown into a tub in such quantity as to fill seven-eighths of its contents. Then a quantity of fresh water is poured on the grain, so as to cover it a few inches. The quality of the water made use of is not indifferent; it is necessary it should be fresh and limpid. The object of this operation being to soften the grain by impregnating all its parts with water, a space of time is requisite proportionate to its dryness and temperature, so that it is less penetrable in winter than in summer, when old than when new; and to regulate the time necessary for this operation, a fixed period should not so much be taken for a basis, as certain signs, easily recognised.

You may always be certain that the corn has been sufficiently steeped when, on being strongly rubbed between the hands, it is completely crushed, without leaving any solid or irreducible particle. All the other means resorted to to recognise the period of its termination are analogous to the latter; such is, for instance, that of cutting it by the nail or crushing it between the teeth. By these means and a little practice, you may always convince yourself that not the least particle of corn has escaped the penetrating and softening action of the water—for this is the only object of steeping. To guide the operator, he is informed that the grain is sufficiently softened and penetrated after having remained from thirty to forty hours in the water, according to the season and the materials made use of.

It is necessary to remark that it is sometimes essential in the heat of the summer to renew the water once or twice, because, without that precaution, a fermentation might take place, which would always prove injurious to future results. When the corn has been sufficiently softened, and is placed under one of the conditions necessary for germination, it is extremely swollen, and increases conspicuously in bulk; this is the reason why it has been recommended not to fill the back entirely.

Then the water is let out of the back through an opening made in the lower part of it, and continues to be drained for ten or twelve hours previously to the succeeding operation, the object of which is to cause the grain to germinate.

The corn, having been suitably steeped in the way above described, is placed on the *malting-floor*, near to which the steeping-back should be placed, to save labour as much as possible. It is placed in a heap on the floor, and left to itself until it becomes palpably warm; this heat is produced by the grain beginning to work, and generally declares itself in from twelve to twenty-four hours after it has been committed to the floor. At this period it is disposed in layers of from 12 to 14 inches in depth, according to the heating state of the floor; they are laid thicker when the temperature of the grain is low, and thinner when more elevated. The influence exercised by the thickness of these layers on the progress of the germination of the corn is very great, with respect to the heat which this thickness may contain; and from this principle it must be concluded that without heat no germination would take place, and that from the moment the watered grain has gained the temperature of 170° to 180°, for instance, it begins to undergo an internal alteration, which produces heat itself. From hence it will be easily conceived that this heat is better retained by a thick layer than by one that is thin; and on this observation is founded the principle which has been emitted on the variation of the thickness to be given to layers on the malt-floor. This kind of fermentation, thus established in grain placed under favourable conditions, soon produces at the end of each grain, and particularly of those that are in the middle of the layer, a white point, which is a sure sign of the commencement of the germination.

This point appears generally twenty-five or thirty hours after the grain has been placed in layers. At this period it is important to turn the grain, so as to place at the bottom of the layer that which was uppermost before; this effect is obtained by removing it to another part of the floor by means of a wooden shovel. It would greatly improve the quality of the malt to submit the grain to this operation once or twice before the appearance of the white point.

The object of this is to regulate the heat of the whole mass, so as to place all the parts of the grain under circumstances equally favourable to germination, and thus to cause the movement to be simultaneous. The heap being thus turned, the white point observed in the grain comes out and presents extricated fibres, which are nothing but the growing roots of the plants. Then it is more important than ever to mind the grain—to remove and turn it frequently, as before recommended, so as to regulate the germination.

This management is essentially necessary, for without it an unequal heat would reign in the mass; this would occasion the roots to grow unequally, and it would be impossible to fix a determinate time for the term of germination. This operation is generally at an end when the fibres have acquired a length of 6 or 7 lines; then the decomposition of the corn is come to a point which is recognised as the most favourable to malt, because at this period the plume which is to form the stalk of the plant is on the point of making its appearance; and if the operation was any longer continued, so as to give this plume the time of shooting out, the malted grain loses a part of the substance useful to the production of spirits. Germination provokes in the corn a change particularly favourable to the success of mashing; it becomes sweetish, and this taste is owing to the saccharification of a small portion of the fecula, or starch. The gluten is partly destroyed, and that which is left becomes soluble in water, from insoluble, which it was before the germination had taken place. By these means the fecula is set at liberty, and the gluten, having become soluble, possesses properties much more energetic than when in its natural state. The object of malting is, then, to convert into sugar a small quantity of the fecula of the corn, and to predispose, at the same time, the rest to a saccharification more complete and prompt, by giving to the gluten the property of being dissolved.

All seasons of the year are not equally favourable to malting; the brewer, whose attention is particularly directed to the malting of his corn for the preparation of beer, prefers the month of March to any other. The grain malted during that month is always of a better quality. Malted corn would not keep in the state of humidity in which it is found on the malt-floor, nor could it be reduced into meal for the purpose of being mashed; it is, then, necessary to *dry* it, which operation is executed as follows:—

The corn, having sufficiently germinated, is taken to the malt-kiln, where it is spread in layers of 8 to 10 inches' thickness; then fire is made under it with combustibles making no smoke, if it can be avoided, because, without this precaution, the malt might contract a smoky smell and taste, which would be transmitted to the spirit. In the brewing of beer various sorts of malt are made use of, which only differ one from another by the temperature they have been submitted to on the kiln; but it is recommended to dry corn destined for distillation at a temperature as near approaching 145° as possible; it is the most favourable to the quality of the malt.

In fact, this temperature, which is also that of mashing, occasions in the wet grain a new formation of sugar, in small quantity, it is true, but this influence of the kiln is not without producing good effect on the subsequent operations, and the temperature of 145° is attended with the greatest success. The combustibles most generally used for the purpose of drying malt are coke or distilled coals; such as that furnished by the establishments of hydrogen gas, or even that proceeding from the distilleries. Next to that comes the vegetable coal, which, if it can be procured at a reasonable price, is very suitable for the purpose.

After this comes the *ash-tree coal*. This species of combustible makes little or no smoke when burning, and exhales sulphurous vapours, which are not at all obnoxious to the quality of the malt. The grain increases greatly in bulk by the operation of mashing; this augmentation may be rated at about one-eighth or ninth part, and their specific gravity decreases in proportion; specific, because the real loss sustained in weight, during the fermentation of grain on the malt-floor, is not easily perceived; but as it occupies after this operation a greater space under the same weight, it is easily conceived that it does not weigh so much under the same bulk.

The reader being now acquainted with the process of malting, and with that of mashing, as also with the effects of these operations, let them now be applied properly. Of all kinds of corn, rye is the one principally used for distillation. Other corn might, it is true, equally be used, but in an economical point of view rye produces the most favourable results. It might be used in the raw state, and might undergo the vinous fermentation, after having been suitably prepared and mashed; but experience has proved the necessity of adjoining to it a certain portion of malted barley. To this effect a quantity of barley is malted, and then mixed to raw rye in the proportion of 20 parts of malted barley to 80 of rye; this mixture is submitted to the operation of grinding and mashing. By this method rye produces more than by any other preparation.

Its produce is greater thus than when used alone, even when malted. Malted barley has the property of rendering rye more fermentable, and it is only in co-operating, by its materials, to the conversion of the fecula of the rye into sugar, during the mashing and even during the fermentation, that it produces this effect. Distillers observe particularly the effect of malted barley on rye in the act of distillation. On consulting them on its mode of acting, they all agree in attributing to it the property of giving lightness to their lob, or paste.

In fact, they have ascertained that, in working with raw corn alone, the fermentation is not so good; and when submitted to distillation, the heavy matter which is found in suspense in the liquid has a very great propensity to precipitate itself to the bottom of the still, and strongly to adhere to it. The least inconvenience attached to this accident is that of communicating a bad taste to the spirit; and it has happened that the bursting of the apparatus has been occasioned by it; so that it is necessary to prevent such accidents, and this result is obtained by mixing the raw grain with a portion of malt.

Distillers explain thus the action of malt, by saying that it gives lightness to their paste, and prevents its falling to the bottom of their still. This explanation, however incomplete it may appear, is nevertheless the expression of a positive fact. Indeed, so long as the fecula has not been converted into sugar, it forms with the water a kind of paste, which has very little fluidity, and which, if exposed to the fire, may easily stick and burn to the bottom of the still.

What happens when malted barley is used with the raw grain? It has already been stated: the barley, by its germination, has undergone a change which renders it more proper to saccharify the fecula. This fact has already been indicated, and still stronger proof will be given when treating of the potato. To saccharify fecula is to destroy the paste which gives viscosity to the liquid, and to supply the fermentation with proper aliments, which is effected, in the distillation of grain, by means of malted barley; and by thus favouring the fermentation a double advantage is obtained—that of having a liquid less heavy, and, of course, more easy to be distilled. It often happens that distillers are in want of malt; then they are forced to distil their raw grain without it.

To obviate a little the inconveniences attached to this way of working, they add, during the mashing, a quantity of chaff. They attribute to this chaff a property analogous to that of malt—that of giving lightness to their matter It has been ascertained that chaff has this property, if not of saccharifying fecula converted into paste, at least to render it fluid, and make it more attackable by the saccharifying agents. It is even probable that barley, besides gluten, contains another matter, which, like chaff, contributes, but more energetically, to the fluidification of fecula. This supposition is the more probable, because no other grain, even when malted, possesses to the same extent the property of saccharifying starch.

For this reason it is always employed, in preference to any other grain, by brewers and distillers. The proportion of chaff used is from 3 to 4 pounds per quintal of raw grain. Its effects are well known in practice, and many distillers add it to their grain even when they use malt. Brewers also make use of chaff, because they have been convinced of its good effects by comparative experiments. The rules which have been laid down before, for the perfect practice of mashing, may be followed without any restriction, whether the corn operated upon be raw or malted, or whether the mixture be composed of grain in those two states.

It may have been remarked that mashing, such as has been described, occasions the fermentable matter to be more or less heavy, according to the quantity of water used, and also according as it has been more or less perfectly executed. Even admitting all the fecula of the corn to have been dissolved during the mashing and fermentation, a certain quantity of husk would always be left in suspension in the liquid, and this quantity is rather large. From this method results, that the distiller is obliged to commit matter to the stills which is very dense and apt to burn, in spite of all precautions that might be taken to prevent this accident.

This method is the only one used in France and Belgium, notwithstanding the inconvenience attached to it. There is another method followed in England and Germany, by which the distillation of pastes, or lobs, is avoided; but whether it necessitates more labour or not, remains a question. Both the French and English methods will be given, and then every one can judge for themselves which possesses the most advantages.

FRENCH METHOD.

Let it be supposed that the quantity of corn made use of is 100 kilogrammes.* This grain, being mixed in the proportion of 80 kilogrammes of rye to 20 of malt, is ground into coarse flour; then deposited, with 2 or 3 kilogrammes of chaff, in a fermenting back containing 12 hectolitres. The steeping is effected by pouring on the meal 3 hectolitres of water at about 110°; then it is mashed with 4 hectolitres of warm and cold water, mixed in such proportion as to give to the mass, after the brewing is over, a temperature of from 145° to 155°.

The tub is covered up, and left to itself for three or four hours. At this period it is filled to within 6 or 8 inches with warm and cold water, mixed in such proportions as to give to the mixture a temperature of about 77°; 1 litre of good yeast is then added. A few hours after the fermentation commences, and proceeds through its various

---

* The French weights and measures are here made use of, as well as in some other parts of this work; their value in English may be ascertained by referring to most any of the arithmetics.

stages in the space of thirty hours; then it is time to commit the liquid to the still. If the operation has been well conducted, from 45 to 50 litres of good spirits at 19° are obtained from 100 kilogrammes of grain. Many distillers are far from producing so much, and there are even some who do not draw more than from 30 to 35 litres. The exiguity of this produce may be the result of several causes; but one of the most influential is the proportion of water used; that is to say, that instead of using 11 hectolitres of water for every 100 kilogrammes of grain, they only use 6. In a continuous work the spent-wash left in the still should be deposited in vats or cisterns constructed for the purpose; there the solid substances will fall to the bottom, and the liquid will remain uppermost.

This liquid may be successfully used in the subsequent operations to dilute the grain after it has been mashed. In this practice is found the advantage of bringing again to fermentation a liquid containing some fermentable substances which have escaped decomposition.

This may be followed up for several successive operations—that is, three, four, and even five; and the grain produces thus as much as 60 litres of spirit of 19° per metrical quintal, produce very considerable, and which could not be obtained by any other means. The use of spent-wash is suspended when, after several successive operations, it is become so sour that instead of offering proper aliments to the fermentation, its acidity would be obnoxious to it. If a smaller proportion of water was used, the same march could not be followed,—at least not to the same extent,—because then the fermentation would require three or four days, instead of thirty hours, and, by these means, cause the spent-wash to be very sour. In this mode, in which the liquid submitted to distillation must necessarily be very heavy, no use can be made of improved apparatuses described elsewhere in this work. In working with this apparatus, care should be taken to stir the first charge submitted to the still until it acquires a temperature approaching that of ebullition, because, without this precaution, the matter might stick and burn at the bottom of the still; this danger disappears when the mass is boiling, and, as in a continuous work the condenser causes the wash to arrive at all times boiling into the still, it will easily be conceived that it is sufficient to agitate the first charge. It would, however, be very advantageous, in this mode of working, to obtain from the grain all the fermentable matter which it contains, and to obtain it in dissolution in water, so as to render the liquid to be submitted to distillation free from husk or any other solid matter. By these means the trouble of agitating the first charge would be avoided; there would be no danger of having the wash burned, or of having bad products; and the various improved apparatuses might be successfully used. No doubt the effects might be obtained by adopting the following method.

ENGLISH METHOD.

It may be stated that this method consists in treating the corn in a double-bottomed tub, and to make the extracts precisely in the same way as the brewers. The grain, composed of malt and rye, being mixed and ground in the same way as for mashing by the French method, 10 kilogrammes of chaff are spread on the first bottom in a layer of 2 centimetres in thickness; 200 kilogrammes of grain are thrown upon it. Then 400 kilogrammes or litres of water, at 35° or 40°, are introduced by a lateral conduit communicating with the empty space between the two bottoms, while the mixture is agitated for five or ten minutes; then the matter is left to subside for a quarter or half an hour, so as to be well penetrated with water. This operation is exactly the same, and its object is the same as that of steeping, which precedes mashing in the method just described.

The only difference existing is in the construction of the apparatus made use of. Immediately after steeping, the matter is again agitated, while 800 kilogrammes of hot water are let into the tub through the same conduit. This time the brewing should last a quarter of an hour, at the end of which the liquid is to be left to itself for at least an hour. At this period the grain is drowned in the water, and a column of liquid tolerably clear covers it; a cock communicating with the space left between the two bottoms is then opened, and as the conical holes of the superior bottom form a species of filtering machine, all the liquid is drained and let into the fermenting backs. After the first extraction, 600 kilogrammes more of boiling water are added in the same way; the mass is again agitated for a quarter of an hour, and left to subside for one hour. The liquid is drained the same as before, to be submitted to fermentation.

The grain on the double bottom has now been sufficiently deprived of all its fermentable substances, which the water has taken away in dissolution in the state of liquid sugar. This operation, which is a true mashing, well understood and well executed, proves beyond doubt the effect of mashing on the corn; it proves that it is, as before remarked, a true saccharification.

When the liquid in the fermenting backs is fallen to a temperature of 75° or 80°, according to the capacity of the tub, yeast is added, and wash without sediment is thus obtained, which can be distilled in all kinds of apparatuses. If the grain left on the double bottom was found not to be sufficiently exhausted, a third extraction might be resorted to. The Germans follow the same method in the distillation of corn, with this difference, that they work with no other grain but what has been malted. Their way of working is then exactly similar to that of English and American brewers, who submit also all the corn they use to the process of malting. To make the best of this method, the proportion of water should be lengthened out with cold water, so as to bring the quantity of water used to ten or twelve times the weight of the corn. Several advantages might be derived from such a proceeding:—1st. A more complete, more rapid, and less acetous fermentation might be thus obtained. 2d. The spent-wash, on leaving the still, might be appropriated to new extracts, and there is no doubt but what greater products would be the result.

FERMENTATION.

Three species of fermentation are recognised — the vinous, the acetous, and the putrefactive; and it has been supposed that these three succeed each other in the order in which they are here called; but it does not follow this rule, as we can see by very slight observation. This important process has been the cause of many contentions in chemistry. Of the vegetable principles, saccharine matter is that which passes with most facility and certainty into the vinous fermentation, and fermented liquors are more or less strong as the juices from which they have been formed have contained a greater or less proportion of sugar before fermentation; for the addition of sugar to the weakly fermentable juices will enable them to produce a strong, full-bodied liquor; and the most essential exit in this process is the disappearance of the sugar, and the consequent production of alcohol. Certain circumstances, however, are necessary to enable it to commence and proceed.

These are—a due degree of dilution in water, a certain temperature, and the presence of substances which appear necessary to favour the subversion of the balance of affinities by which the principles of the saccharine matter would otherwise be retained in union, or, at least, would be prevented from entering into those combinations necessary to form vinous spirit.

These substances, from this operation, are named ferments. First, a certain proportion of water to the matter susceptible of fermentation is requisite. If the latter is in large quantity, proportioned to the water, the fermentation does not commence easily or proceed so quickly; on the other hand, too large a proportion of water is injurious, as causing the fermented liquor to pass speedily into the acetous fermentation. The necessary consistence exists naturally in the juice of grapes and in the saccharine sap of many trees, and other spontaneously fermentable liquors; for if these very liquors be deprived by gentle evaporation of a considerable portion of their water, the residue will not ferment until the requisite consistence is restored by the addition of a fresh portion of water.

Secondly, a certain temperature is not less essential; it requires to be at least 55° of Fahr. At a temperature lower than this, fermentation scarcely commences, or, if it has begun, proceeds very slowly; and, if too high, requires to be checked, to prevent it from passing into the acetous state.

Lastly, though sugar or substances analogous to it are the matters which serve as the basis of fermentation, and from which its products are formed, the presence of other matter is requisite to the process. It has been often stated that sugar alone, dissolved in a certain quantity of water, and placed in a certain temperature, will pass into a state of fermentation.

It is, however, doubtful if this happens with a solution of *pure* sugar, and any change which is observed is imperfect and irregular; nor does the liquor become vinous, but rather sour. The substance usually added to produce fermentation is called yeast. When the proper sort of ferment is pitched upon, the operator is next to consider its quantity, quality, and manner of application. The quantity must be proportioned to that of the liquor, to its tenacity, and the degree of flavour it is intended to give, and to the despatch required in the operation. From these considerations he will be enabled to form a rule to himself, in order to the forming of which a proper trial will be necessary to show how much suffices for the purpose. The greatest circumspection and care are necessary in regard to the quality of the ferment, if a pure and well-flavoured spirit be required.

It must be chosen perfectly sweet and fresh, for all ferments are liable to grow musty and corrupt; and if in this state they are mixed with the fermentable liquor, they will communicate their nauseous and filthy flavour to the spirit, which will scarcely ever be got rid of by any subsequent process. If the ferment be sour, it must by no means be used with any liquor, for it will communicate its flavour to the whole, and even prevent its rising to a head, and give it an acetous instead of a vinous tendency. When the property of well-conditioned ferment is prepared, it should be diffused in the liquor to be fermented in a tepid or lukewarm state. When the whole is thus set to work, secured in a proper degree of warmth, and kept from a too free intercourse with the external air, it becomes, as it were, the sole business of nature to finish the operation, and render the liquor fit for the still.

The first signs of fermentation are—a gentle intestine motion, the rising of small bubbles to the top of the liquor, and a whitish, turbid appearance. This is soon followed by the collection of a froth or head, consisting of a multitude of air-bubbles entangled in the liquor, which, as the process advances, rise slowly to a considerable height, forming a white, dense, permanent froth

A very large portion of the gas also escapes, which has a strong, penetrating, agreeable vinous odour. The temperature of the liquor at the same time increases several degrees, and continues so during the whole process.

Sooner or later, these appearances gradually subside; the head of the foam settles down, and the liquor appears much clearer and nearly at rest, having deposited a copious sediment, and, from being viscid and saccharine, is now become vinous, intoxicating, much thinner, or of less specific gravity. The process of fermentation, however, does not terminate suddenly, but goes off more or less gradually, according to the heat at which it was commenced, and of the temperature of the external air. The gas of fermenting liquors has long been known to consist for the most part of carbonic acid; it will therefore extinguish a candle, destroy animal life, convert caustic alkalies into carbonates, and render lime-water turbid by recomposing limestone, which is insoluble, from the quicklime held in solution. The attenuation of liquors, or the diminution of their specific gravity by fermentation, is very striking. This is shown by the hydrometer, which swims much deeper in fermented liquor than in the same materials before fermentation.

No doubt much of this attenuation is owing to the destruction of the sugar, which dissolves in water, adds to its density, and to the consequent production of alcohol, which, on the contrary, by mixing with water, diminishes the density of the compound. The tract or mucilage also appears to be in some degree destroyed by fermentation, for the gelatinous consistence of thick liquors is much lessened by this process; the destruction of this principle, however, is by no means so complete as of the sugar, many of the full-bodied ales, for example, retaining much of their clamminess and gelatinous density, even after having undergone a very perfect fermentation.

Atmospheric air, it seems, has a no less share whatever in vinous fermentation; for it will take place full as well in closed as in open vessels, provided space is allowed for the expansion of the materials and the copious production of gas. The great question to be determined is, What may be the substance or circumstance which disposes sugar to ferment? for it has been proved that sugar will not of itself begin this spontaneous change into carbonic acid and alcohol, though when once begun the process will probably go on without further assistance. Some of the most common fermenting ingredients, as the sweet infusion of malt, technically called wort, it is well known, will slowly enter into fermentation without the addition of yeast; hence chemists have sought in this substance for the principle which gives the first impulse to the fermentation of sugar.

Generally, it has been supposed that no substance enters into the vinous fermentation except sugar, or from which sugar may be extracted, and that the process of malting grain was necessary to develop the sugar or saccharine matter, to render it susceptible of vinous fermentation. The practice, however, of grain distillers proves this to be a mistake, as they obtain as much spirit from a mixture of malted barley with unmalted grain as if the whole were malted. The properties of the fermented liquor, its odour, pungency, and intoxicating quality, are owing to the presence of a substance which can be separated from it by distillation, and which in a pure state possesses these qualities in a much higher degree.

It constitutes, in the state of dilution in which it is obtained by distillation, vinous spirit, or, as obtained from the different fermented liquors from which it derives peculiarities of taste and flavour, the spirituous liquors of commerce. These, by certain processes, afford this principle pure, and the same from all of them; in this pure state it is called spirits of wine, or alcohol.

RECTIFICATION.

To obtain a pure, clean, flavourless spirit, no attempts have been wanting on the part of the most diligent rectifiers. It has long since been observed that rectification is an operation performed in various ways, some of which scarcely deserve the name; because, instead of freeing the spirit from its gross essential oil and phlegm, they alter the natural flavor of that which comes over in the process. The principal business is to separate the spirit from the essential oil of the malt, &c. In order to do this, care should be taken in the first distillation that the spirit, especially from malt or grain, should be drawn by a gentle fire, by which means a great part of the essential oil will be kept from mixing with the spirit; for experience daily proves that it is much easier to keep asunder than to separate subjects once mixed. In order to rectify low wines, they should be put into a tall body or alembic, and gently distilled in *balnea mariæ*; by this means a large proportion, both of oil and phlegm, will remain in the body.

But if, after this operation, the spirit should be found to have too much of essential oil, it must be let down with fair water, and gently distilled; by this it may be brought to any degree of purity. The redundant oil may, however, be separated from proof spirit, &c. by the method already proposed, especially if it be previously filtrated through paper, thick flannel, sand, stone, &c. placing at the bottom of each some cotton-wool, for absorbing the oil that escapes the filter. But the slowness of this operation has caused many distillers to substitute caustic alkalies, which only destroy the natural flavour of the spirit.

In fact, almost every distiller pretends to have some secret nostrum for rectifying his spirits; however, they are all reducible to three, namely: by fixed alkaline salts, by acid spirits mixed with alkaline salts, and by saline bodies and flavouring additions. Some distillers use quicklime in rectifying their malt spirit, which cleanses it considerably; but if chalk, calcined and well-purified animal bones, were substituted for quicklime, the spirit would have a less alkaline or nitrous flavour, and consequently the flavouring ingredients might be added to it with more success than by the other methods. Neutral salts and soluble tartar might also be used; but fine dry sugar seems best adapted for the purpose of rectifying these spirits, as it readily unites with the essential oil, detains, and fixes it, without imparting any urinous, alkaline, or other nauseous flavour to the spirits.

COMMON PROCESS OF MALT DISTILLING

Take 60 quarters of barley grist, ground low, and 30 quarters of pale malt, ground rather coarse; make your lob with 10 quarters of the malt, ground into coarse flour, and 30 barrels of liquor, at the heat of 170°. Row or blend them into a uniform mass, and mix them thoroughly with the major part of the first wort, and pump them up together into the coolers. When cooled to the temperature of 55°, they are to be let down into the fermenting-back, to the reserved part of the first worts; say, 30 barrels previously pitched at 60°, with 10 stone of fresh porter yeast, which, with the rest of the worts at 55°, altogether compose a back of distillers' wash. Take the specific gravity of the worts previous to their descent into the backs, and before any yeast is added, and note it down in a book or table prepared for that purpose; do this every twelve hours for three or four days, during which it may be found to increase in gravity and sweetness, from the augmenting force of the fermentation, resolving the gluten and extracting the saccharine matter. This is malting in the gyle-tun, or fermenting-back. When the gravity seems to be stationary, or rather decreasing, a vinous tartness will begin to succeed the previous sweetness, the fermentation becomes more vigorous, and the gravity more rapidly decreases; before it arrives at this period, a sensible decrease of gravity, and conspicuous change of flavour from sweet to tart, usually take place.

Closely observe every change and appearance in the fermentation, and note it down in your book. In the course of twelve or fourteen days, the yeast-head will fall quite flat, which denotes the fermentation being nearly over. If the heat appears by the thermometer to drop, and the fermentation has gone on well, or if the attenuation appears by the hydrometer to have reduced the gravity of the wash from its original weight of 28, 30, or greater number of pounds, 2, 3, or 4 pounds per barrel, and the wash should have a vinous odour and flavour, then all is right. At this period some add 20 pounds of common salt and 30 pounds of flour; rouse and keep the fermenting-back close, as it should have been during the whole process.

In three or four days it will taste quite tart, and should be immediately distilled. The wash, duly fermented, is committed to the still; all the time it is running in, it should be roused up or agitated in the fermenting-back by a stirring-engine, to mix the thick and thin parts together into one mass, and enable it to be sufficiently fluid to flow into the still, where it is kept fluid by the stirring-engine of the still until it boils, when the agitation of the boiling usually keeps it from burning and giving empyreumatic or burnt flavour to the low wines; which taint will inevitably rise from the low wines in the spirit-still during the doubling or distilling the spirits of the second extraction. This spirit is usually sold by weight, delivered to rectifying distillers at one to ten over proof, who rectify or distil it over again, combining it with certain ingredients in order to clarify it from its gross oil and other impurities, with the view to render it fit for making gin, brandy, rum, and fine cordial compounds, &c. as the case may be.

FRENCH PROCESS OF DISTILLING AND PREPARING BRANDY.

This process differs in nothing from the ordinary process practiced in England and this country, in the same manner as from malt-wash or molasses. The French only observe, more particularly, to throw a little of the natural lees into the still along with the wine, because they find this gives their spirit the flavour for which it is so much admired But though brandy is extracted from wine, experience tells us that there is great difference in the grapes from which the wine is made. Every soil, every climate, every kind of grapes vary with regard to the quality or quantity of spirits extracted from them.

Some grapes are only fit for eating; others for drying, as those of Damascus, Corinth, Provence, and Avignon, but not fit to make wine. Some wines are proper for distillation, others less so. Those of Languedoc and Provence afford a great deal of brandy by distillation, when the operation is made in their full strength; the Orleans wine and those of Blois afford still more. The best wines are those of Cognac and Audaye, which, however, are among those that are least drunk in France; whereas those of Burgundy and Champagne, though of a very fine flavour, yield but very little in distillation.

It may also be further observed, that all the wines for distillation, as those of Spain, the Canaries, of Alicant, Cyprus, St. Peres, Toquet, Graves, Hungary, and others, yield very little brandy by distillation, and consequently would cost the distiller considerably more than he could sell it for. What is drawn from them, however, is good, always retaining their saccharine quality and rich flavour; but, as it grows old, this flavour often becomes aromatic, and is not agreeable to all palates. Hence brandies differ as they are extracted from different sorts of grapes; nor would there be so great a similarity as there is between the different kinds of French brandies, were the strongest wines used for distillation. But this is rarely the case: the weakest and lowest flavoured wines only are drawn for their spirit, or such as prove absolutely unfit for any other use. A large quantity of brandy is distilled in France during the time of the vintage; for all those poor grapes that prove unfit for wine are usually first gathered, pressed, their juices fermented, and directly distilled.

This rids their hands of the poor grapes at once, and leaves their casks empty for the reception of better. It is a general rule in France not to distil any wine that will bring a good price as wine; for in this state the profits upon them are much higher than when reduced to brandies. The large stock of small wines with which they are almost overrun in France sufficiently accounts for their making such quantities of brandy—more than in any other country which has a warmer climate, and is better adapted to the production of grapes.

Nor is this the only fund for French brandies; for all the wines that turn sour or sharp are condemned to the still; and all such as they can neither export nor consume at home, which amounts to a large quantity, as much of that laid in for their families is often so poor as not to keep from one season to another. Hence many American and English spirits, with proper management, are convertible into brandies that in many respects, provided the operation be neatly performed, can scarcely be distinguished from the French. Even a cider spirit and a crab spirit may, from the extraction, be made to resemble the fine and thin brandies of France. The art of colouring spirits owes its rise to observations on French brandies, and being found to have been derived from the oak of the cask, it is no difficulty to imitate it to perfection.

METHOD OF PREVENTING THE DETERIORATION OF BRANDIES.

It is certain that when brandy is kept in vessels the pores of which will not admit of any transmission of the liquor, (as glass, for instance,) the brandy will improve, instead of getting worse. The wine-merchant has no idea of bottling off a whole store; but, without much expense, he may render the hogshead absolutely impermeable, and besides, the expense in doing this, being once undertaken, will be available for a considerable time.

To effect this, a very large tub should be well hooped; two layers of oil colours then being laid on, these should be followed by a good coating of pitch and tar: this will put every idea of evaporation out of the question. In a barrel thus treated, the spirit of the brandy may be preserved three years without the least loss, either in quantity or quality. These large tubs or reservoirs, being built into the brickwork, &c. of the storehouse, may serve during a considerable lapse of time without reparation.

OF MALT WHISKY.

In this country the term distillation is often applied to the whole process of converting malt or other saccharine matter into spirits or alcohol. In making malt whisky, 1 part of bruised malt, with from 4 to 9 parts of barley-meal, and a proportion of seeds of oats corresponding to that of the raw grain, are infused in a mash-tun of cast-iron, with from 12 to 13 gallons of water, at 150° Fahr., for every bushel of the mixed farinaceous matter. The agitation then given by manual labour or machinery, to break down and equally to diffuse the lumps of meal, constitutes the process of mashing. This operation continues two hours or upward, according to the proportion of unmalted barley; during which the temperature is kept up by the affusion of 7 or 8 additional gallons of water a few degrees under the boiling temperature. The infusion, termed wort, having become progressively sweeter, is allowed to settle for two hours, and is run off from the top to the amount of one-third of the bulk of the water employed.

About 8 gallons more of water, a little under 200° Fahr., are now admitted to the residuum, infused for nearly half an hour, with agitation, and then left to subside for nearly an hour and a half, when it is drawn off. Sometimes a third affusion of boiling water, equal to the first quantity, is made, and this infusion is generally reserved to be poured on the new farinæ; or it is concentrated by boiling, and added to the former liquors. To prevent acetification, it is necessary to cool the worts down to the proper fermenting temperature of 65° or 70° as rapidly as possible.

Hence they are pumped immediately from the mash-tun into the extensive wooden troughs, 2 or 3 inches deep, exposed in open sheds to the cool air; or they are made to traverse the convolutions of a pipe immersed in cold water. The wort being now run into a fermenting-tun, yeast is introduced, and added in nearly equal successive portions during the three days, amounting in all to about 1 gallon for every 2 bushels of farinaceous matter. The temperature rises in three or four days to its maximum of 80°; and at the end of eight or twelve days the fermentation is completed, the tuns being closed up during the last half of the period.

The distillers do not collect the yeast from their fermenting-tuns, but allow it to fall down, on the supposition that it enhances the quantity of alcohol. Quick distillation does not injure the flavour of spirits—this depending almost entirely upon the mode of conducting the previous fermentation. In distilling off the spirit from the fermented wort or wash, an hydrometer is used to ascertain its progressive diminution of strength; and when it acquires a certain weakness the process is stopped, by opening the stopcock of the pipe which issues from the bottom of the still, and the spent-wash is removed. There is generally introduced into the still a piece of soap, whose oily principle, spreading on the surface of the boiling liquor, breaks the larger bubbles, and of course checks the tendency to froth up.

Indian corn, in this process, can be used instead of the barley, and the raw oats can be omitted.

PROCESS FOR MAKING DUTCH GENEVA.

For a long time the Dutch have been famous for the manufacture of an excellent kind of spirits, called Hollands. They mash 112 pounds of barley and 228 pounds of rye-meal together, with 460 gallons of water of the temperature of 162°. After the infusion has stood a sufficient time, cold water is added till the strength of the wort is reduced to 45 pounds per barrel. The whole is then put into a fermenting-back, at the temperature of 80°; the vessel is capable of holding 500 gallons. Half a gallon of yeast is added; the temperature rises to 90°, and the fermentation is over in forty-eight hours. The attenuation is such, that the strength of the wash is not reduced lower than 15 pounds per barrel.

There is another method given for making Dutch geneva, which is as follows:—1 hundred-weight of barley-malt and 2 of rye-meal are mashed with 460 gallons of water, heated to 162° Fahr. After the farinæ have been infused for a sufficient time, cold water is to be added till the wort becomes equivalent to 45 pounds of saccharine matter per barrel. Into a vessel of 500 gallons capacity the wort is now to be put, at a temperature of 80°, with half a gallon of yeast.

The fermentation instantly begins, and is finished in forty-eight hours, during which the heat rises to 90°. The wash, not reduced lower than 12 or 15 pounds per barrel, is put into the still along with the grains. Three distillations are required; and at the last a few juniper-berries and hops are introduced, to communicate a flavour.

The attenuation of 45 pounds in the wort to only 15 in the wash, shows that the fermentation is here very imperfect and uneconomical; as, indeed, might be inferred, from the small proportion of yeast and the precipitancy of the process of fermentation.

On the other hand, the very large proportion of the porter-yeast, in a corrupting state, used by the Scotch distillers, cannot fail to injure the flavour of their spirits. The finest Hollands geneva is said to be made in Holland from a spirit drawn from wheat, mixed with a third or fourth part of malted barley, and twice rectified over juniper-berries; but, in general, rye-meal is used instead of wheat. They pay so much regard to the water employed, that many distillers send vessels to the Meuse on purpose to bring it; but all use the softest and clearest river-water they can get.

In England, it is the common practice to add oil of turpentine, in the proportion of 2 ounces to 10 gallons of raw spirit, with 3 handfuls of bay-salt; and these to be drawn off till the feints begin to rise. Corn, or spirit of molasses, is also flavoured by a variety of aromatics, with or without sugar, to please different palates; all of which are included under the technical term of "compounds," or "cordials."

Other articles have been employed for the fabrication of spirit, though not commonly; for instance, carrots and potatoes. To obtain pure alcohol, different processes have been recommended; but the purest rectified spirit, obtained as above described, being least contaminated with foreign matter, should be preferred. Some recommend the drawing off half the spirit in a water-bath; then to rectify this twice more, drawing off two-thirds each time; to add water to this alcohol, which will turn it milky, by separating the essential oil remaining in it; to distil the spirit from this water, and finally rectify it by one or more distillations. Others set apart the first running, when about one-fourth is come over, and continue the distillation till they have drawn off about as much more, or till the liquor runs off milky.

The last running they put into the still again, and mix the first half of what comes over with the preceding first product. This process they repeat again, when all the first products mixed together are distilled afresh. When half of the liquor has come over, this is to be set apart as pure alcohol. Alcohol in this state, notwithstanding, is not so pure as when it has been dephlegmed, or still further freed from water, by means of some alkaline salt. Muriate of soda has been recommended for this purpose, deprived of its water of crystallization by heat, and added hot to the spirit; but the subcarbonate of potash is preferable.

About a third of the weight of the alcohol should be added to it in a glass vessel, well shaken and then suffered to subside. The salt will be moistened by the water absorbed from the alcohol, which being decanted, more of the salt is to be added; and this is to be continued till the salt falls dry to the bottom of the vessel. The alcohol, in this state, will be reddened by a portion of pure potash which it will hold in solution, from which it must be freed by distillation in a water-bath. Dry muriate of lime may be substituted advantageously for the alkali. Lastly, as alcohol is much lighter than water, its specific gravity is adopted as the test of its purity. Fourcroy considers it as rectified to the highest point when its specific gravity is 829, that of water being 1000; and perhaps this is as far as it can be carried by the best process.

PROCESS FOR BREWING HOLLANDS GIN.

Their grist is composed of 10 quarters of malt, ground considerably finer than ordinary malt, distillers' barley-grist, and 3 quarters of rye-meal; or, more frequently, of 10 quarters of rye and 3 quarters of malt-meal. The 10 quarters are first mashed with the least quantity of cold water it is possible to blend it with; when uniformly incorporated, as much boiling water is added as forms it into a thin batter. It is then put into one, two, or more casks, or gyle-tuns, with a much less quantity of yeast than is usually employed by distillers.

Generally, on the third day they add the malt or rye-meal, previously made into a kind of lob, prepared in a similar manner, except in not being so much diluted; but not before it comes to the temperature of the fermenting-wash; at the same time adding full as much yeast as when at first setting the backs. The principal secret in the management of the mashing part of the business is, in first thoroughly mixing the malt with the cold water, that it may still remain sufficiently thin after the addition of the fine meal under the form of lob; and in well rousing all together in the back, that the wash may be sufficiently diluted for distilling, without endangering its burning to the bottom of the still. Thus they commodiously reduce the business of brewing and fermenting to one operation.

By using cold water, uniformly, to wet the malt, all danger of clogging the spending of the tap would necessarily be avoided; but here there is no occasion to do any thing more than to dilute the wash, consisting of the whole of the grain, thin enough to be fermented and distilled together, by which means the spirit of the bran and husky part, as well as of the flour, are completely extracted. Yet this wash, compared to the ordinary distillers' wash of this country and England, is about three-eighths thinner. For these reasons, they obtain more spirit from their grain, and of a better quality, with not half the trouble taken by other distillers.

Their backs usually contain as much wash as serves for one distillation. The gravity of the distillers' wash at Weesoppe, in the neighbourhood of Amsterdam, is but 18 pounds per barrel—very little more than half the gravity of that of the English distillers. Their stills usually hold from 300 to 500 gallons each; they constantly draw off 3 cans of phlegm after the runnings cease to form on the head of the still, when distilling wash, and 5 cans when distilling low wines; a practice not followed elsewhere.

This, and the great quantity of rye they use, causes their spirit to be much more acid; and the diluteness of their wash is a very good reason for the greater purity of their spirit, though most writers contend that it is not so clear.

PROCESS FOR RECTIFICATION INTO HOLLANDS GIN.

This process is conducted as follows:—To every 20 gallons of spirits of the second extraction, about the strength of proof spirit, take 3 pounds of juniper-berries, and 2 ounces of the oil of juniper, and distil with a slow fire until the feints begin to rise; then change the receiving-can: this produces the best Rotterdam gin. An inferior kind is made with a still less proportion of berries, sweet fennel-seeds, and Strasburg turpentine, without a drop of juniper-oil. It, and a better sort, but inferior to the Rotterdam gin, are made at Weesoppe. The distillers' wash at Schiedam and Rotterdam are still lighter than that at Weesoppe.

Strasburg turpentine is of a light yellowish-brown colour, and very fragrant, agreeable smell; its taste is the bitterest yet the least acid of the turpentines. The juniper-berries are so very cheap in Holland, that they must have more reasons than mere cheapness for being so much more sparing of their consumption than distillers in this country. Indeed, they are not in the habit of wasting any thing. The two principal modes of preparing geneva in Holland have thus been described by an eminent distiller:—

"A quantity of flour of rye, coarsely ground, is mixed with a third or fourth part of barley-malt, proportioned to the size of the tub in which the vinous fermentation is to be effected.

"This they mix with cold water, and then stir it with the hand, to prevent the flour from gathering into lumps, and to facilitate its dissolution. When this point is attained, water is added of the heat of the human blood— about 98°. The whole is well stirred, after which the ferment is mixed with the wort, being previously diluted with a little of the liquor.

"The fermentation generally begins six hours afterward. If it commences earlier, there is reason to apprehend it will be too strong, and means are employed to check it. If the fermentation be well conducted, it generally terminates on the third day, when the liquor grows transparent, and assumes an acrid taste, hot and biting on the tongue. The wash is then well roused, and the mash, with all the corn, is put into the still; and then commences the first distillation, which is conducted very slowly, which is a matter of the utmost importance. This liquor is then rectified over juniper-berries once or twice, according to the sort of spirit which it is intended to produce. For common use one rectification is deemed suficient, though it is not considered so fine, pleasant, or delicate as that which has undergone several rectifications, and which is called double geneva.

"Some distillers mix the juniper-berries with the wort, and ferment them together; but in that case they only draw a spirit from it for the interior or for exportation: the juniper, however, is most commonly used at the rectification, and not before. In the second method pursued by the best distillers, the malt and rye are mixed with warm water in given proportions, and thoroughly blended together until all the farinaceous substance is incorporated;

the liquid is then allowed to rest until the flour has settled at the bottom.

"The wort is afterward permitted to flow into the fermenting-tun, where a similar operation takes place with another quanity of water poured upon the same grain; and these operations are repeated until the wort thus drawn from it at different times has abstracted the whole saccharine matter in the flour. This liquid is put into the fermenting-tun or vessel, and when it comes to the proper temperature, about blood-heat, the ferment, or yeast, is added. The fermentation is considered more mild and regular by this method than the other. Some pour all the water they intend to use into the tub or kieve at once, and put the flour gently into it, while two or more persons are employed in stirring it with sticks made for that purpose, to mix the flour, and to prevent it from gathering into lumps. When the whole of it is properly reduced and mixed together, they proceed to draw it off into a cooler, before it is put into the fermenting vessel.

"In all cases the gravity of the worts is low, seldom exceeding 45; and, by distilling from a mixture of wash and grains, the produce is allowed to be much greater than that obtained in Great Britain from potato wash."

DISTILLATION OF COMMON GIN.

TAKE of ordinary malt spirits 10 gallons; oil of turpentine, 2 ounces; juniper-berries, 1 pound; sweet-fennel and caraway seeds, of each 4 handfuls; bay-salt, 3 handfuls. Draw off by a gentle fire till the feints begin to rise, and make up your goods to the strength required; say, 10 gallons of spirit will make about 15 gallons of common gin.

SPIRIT OF POTATOES.

IN selecting potatoes for distillation, those that are the most farinaceous when boiled, and the most agreeable to the palate, must always be preferred to any others. The most favourable season for distilling potatoes is from the month of October, when they are harvested, to the month of March, when they begin to germinate. The latter circumstance has great influence on their quality; it causes their proportion of fecula to decrease, and renders their spirituous produce much less in quantity. As the distillation of potatoes more especially takes place in the winter and in the latter part of the year, the frost which comes almost regularly at that time might injure the quality of the potatoes, if proper precaution was not taken to protect them against its influence.

To this effect, it is necessary to place them in warehouses, or other suitable places, where the temperature never gets so low as to endanger them. Cellars are very proper to fulfil this object, because they keep almost invariably, in winter as in summer, at a temperature of 55° Fahr. There are two methods of preparing potatoes for fermentation; the object of both is to saccharify their fecula. The first is by means of malted barley or Indian corn, and, though practised in town distilleries, is more generally followed in the country, because it is more intimately connected with the feeding of cattle; and is composed of three operations. Potatoes were first used for distillation many years ago, and the method then adopted consisted in submitting them to the action of boiling water, as it is still done in their preparation for food.

For this purpose stills of 3 or 4 hectolitres were used, but the difficulty of getting the potatoes out of those vessels, and the expense of fuel, soon caused it to be replaced by that of steaming them, which mode is much less expensive. Various apparatuses have been invented for boiling potatoes; the following is said to be the most perfect:—

Fig. 13 represents this apparatus erected on brick-work.

A is a copper still, provided with a cover B strongly fastened to the neck of the still by means of iron nippers, similar to those made use of in the improved apparatus of Adam and Bérard, which will be spoken of hereafter

The cover B bears a curved tube C D, which carries off the steam; the extremity D of this tube is furnished with a collar, by which means it may be screwed to any other tube: this mode of fastening has been considered the most commodious. E F is a safety-tube, which is also used in filling the still, and which plunges into the latter to about 5 centimetres from the bottom.

This tube is a safety-tube, insomuch as it would cause the water to run out by its orifice E if the pressure in the still was too great; and it also shows, by giving passage to the steam, when the water should be renewed in the latter.

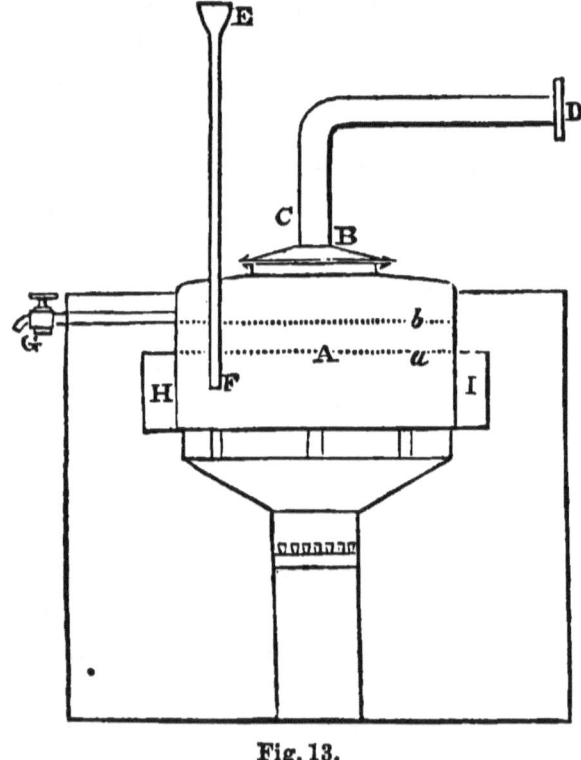

Fig. 13.

G is a cock fixing the level of the water in the still when charged. To effect this it is always open, in a continuous work, during the filling of the still. The function of this cock is twofold: it gives access to the air, and thus prevents the dangers from absorption. The still A is calculated to produce 168 pounds of steam per hour, provided it be supplied with boiling water; for if this was to be brought to ebullition by the still itself, a much less quantity of steam would be formed. It contains 230 litres up to the line a, and 306 up to b, which is the full charge. It would be easy to supply this still constantly with boiling water; this might be done by placing above it a small vessel filled with water, which might be brought to ebullition by means of the heat which escapes from the first still: a great economy in combustibles might thus be obtained.

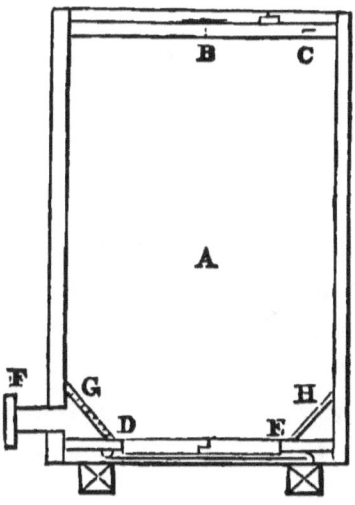

Fig. 14.

The tube D, adapted to another tube F in fig. 14, conveys the steam produced in this still into a tub (fig. 14) calculated to hold 1280 litres of matter; but, as it is necessary never to fill it entirely with raw potatoes, because ebullition causes these to swell, it is only charged with 11 hectolitres.

A (fig. 14, on the preceding page) is a cylindrical tub, made of strong oak; the interior of this tub is lined with copper or with lead, so as to render it sufficiently solid.

The potatoes are introduced through a trap B C fixed in the head of the tub, and they are discharged through a double trap D E placed in the bottom.

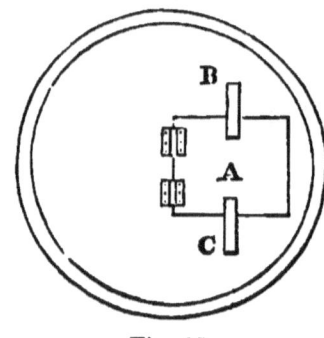

Fig. 15.

Fig. 15 represents the head of this tub.

A is a trap, hanging on two hinges, which can be shut and closely maintained to the head of the tub by means of two buttons, B and C.

Fig. 16 represents the bottom of the tub.

A and B form a double trap, opening in the middle, and hanging on four hinges a, b, c, d. It is opened by removing an iron bar C E, fastened by the end C to an iron cramp by means of a pink, which gives it sufficient play.

This bar slips into E, where it is retained by means of a button D; and when it is unhooked and separated from

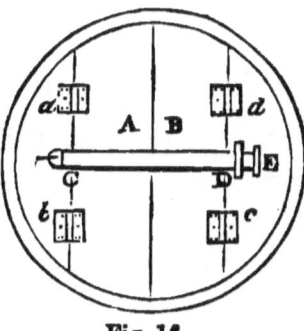

Fig. 16.

E, it hangs by the extremity C, and causes the double trap to open, and thus to let the boiled potatoes fall out. The tube F (fig. 14) is destined to introduce the steam into the tub A; its internal orifice is protected by an inclined plane c, full of holes. This keeps the potatoes from stopping the tube F, and from thus becoming an obstacle to the free entrance of the steam; it is also useful, the same as the plane H, to prevent the boiled potatoes from being left in the corners of the tub. By means of this apparatus and of the boiler just described, 900 kilogrammes of potatoes may be boiled in one hour; this will require 14 kilogrammes of coals. It is necessary, as a matter of economy, to lute the various parts of this apparatus through the joints of which steam might be lost. This is done by means of clay mixed with some other substance. The most convenient place that can be given to the tub is above the hopper of the reducing machine, which will shortly be spoken of.

REDUCTION OF POTATOES.

When the potatoes have been boiled suitably, they occupy a greater space, and this dilatation causes their peel, which is not very strong, to be broken; this renders them proper to be submitted to the action of the reducing machine, of which a description will now be given.

This machine is represented by figs. 17 and 18.

Fig. 17 is a lateral elevation of it, showing the side of the machine where the two wheels work in each other.

Fig. 18 is a plan of the machine, as seen from above, without the hopper. The same letters represent the same pieces in both figures.

A A A A is a strong frame built in oak.

B and C are two cylinders made of wood or of stone, but better of cast-metal. These two cylinders are of equal diameters; they are a short distance from each other, and their function is to break the potatoes by their unequal rotatory motion. This unequal rotation is obtained by means of two wheels of different diameters, a and b.

D is a movable hopper, supported by the frame in $c\ d$. It is made for the purpose of receiving the potatoes to submit them to the action of the cylinders.

E, F are two handles fixed on the axis of the cylinder c. They serve to work the machine.

G H is a scraper. It hangs on e, and the weight g, which acts upon the lever G e, presses the extremity H of the scraper against the cylinder B. This scraper

POTATOES—REDUCTION.

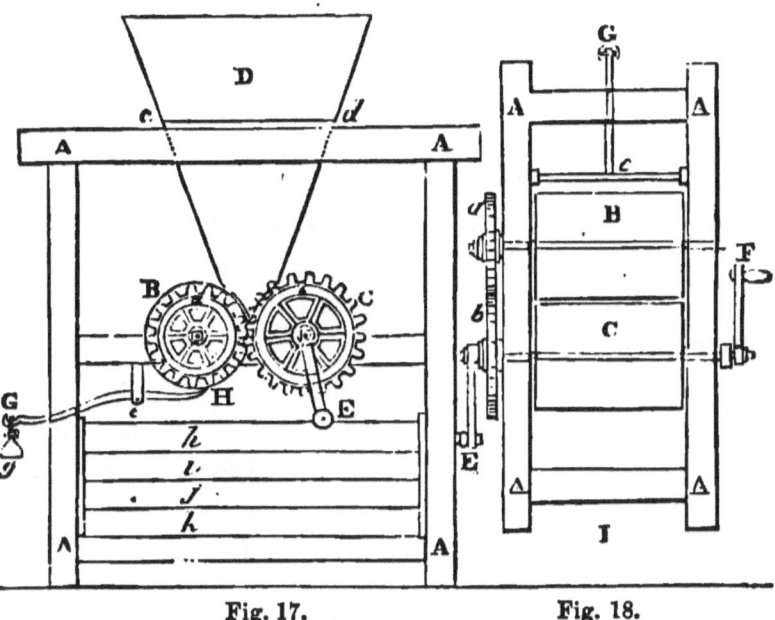

Fig. 17.   Fig. 18.

serves to separate from the cylinder the broken potatoes that might stick to it.

h, i, j, k are movable boards held by two grooves; these close the interior part of the machine, and form one side of a chest which receives the broken potatoes. These are let out on the side I through a mobile shutter.

The construction of this machine is extremely simple, not at all expensive, and very little subject to repairs. Two handles have been fixed to it, though one man might work it; but it would be better, in an uninterrupted work, to employ two men. 1000 kilogrammes (about 2000 pounds) of potatoes may thus be reduced in the course of ten or twelve hours; a greater product might be obtained by applying more force to it, and by making use of such wheels as would accelerate the rotation of the cylinders.

MASHING OF POTATOES.

The potatoes, having been reduced into paste, are mashed with malted barley or Indian corn, at a temperature of 170° or 180°. Suppose a quantity of potatoes are to be worked sufficient to produce 12 hectolitres of fermentable matter. A tub containing at least 13 hectolitres is made use of; the pulp of 400 kilogrammes of potatoes is deposited in it in the state of paste. About 25 kilogrammes of malted barley or Indian corn, as the case may be, are added to this paste, together with a quantity of warm and cold water sufficient to establish in the tub a temperature of from 70° to 75° or 80° Fahr., which is the heat requisite for the steeping of grain; the mixture is strongly agitated, and left to subside for one-quarter of an hour, or perhaps half an hour.

Then, while the mass is again stirred, boiling water is introduced until the thermometer has risen to 172° or 180°. The paste is now left to macerate for two or three hours; then a mixture of cold and warm water is added, in such proportion as to form 12 hectolitres of liquid at 80°. 1 litre, or 2 wine-pints, of good yeast is then added, and the fermentation is established in a few hours. It is worthy of remark that in this case, as also in the mashing of corn, the saccharifying action of the barley, signalized in the mashing of rice, is very little perceived.

After the last mashing, there is only a small quantity of sweet liquid swimming above the paste, so that in this state the fecula of the potatoes has not been entirely converted into sugar. The reason is, that the fecula has not been sufficiently decomposed in the potatoes boiled by steam. To liquefy and convert this fecula into sugar by means of malted barley, it is necessary to place it in immediate contact with the latter; the husks, and the granular and crystalline state in which the fecula is found in potatoes boiled by steam, fulfil but imperfectly the condition necessary for the complete saccharification which has been observed in the mashing of rice.

This saccharification is completed in the fermentation, at a much lower temperature, it is true, but not in so short a period. In fact, this conversion of fecula into sugar takes place as long as the fermentation lasts. To this process, simple in itself, are attached all the inconveniences inherent to the distillation of pastes. Agriculture, which is deeply interested in this kind of industry, has long since been in want of means to remove this imperfection, without too hasty innovations, and without affecting the simplicity and economy belonging to this method.

There are some important modifications to the method just detailed, which it may not be amiss to give while on this subject.

These modifications may be executed in two ways: the first consists in isolating the fecula of the potato, to work it with malted barley or Indian corn; by the other the separation of the fecula is avoided, by directly operating on potatoes simply divided by the rasp. As the fabrication of fecula will be useful to the reader, not only in this, but in the second method, a description will be given of it, such as is practised in Paris for the purpose of distillation. This restriction is established for the preparation of fecula, because it does not require, for distillation, the same care and management as when made for domestic purposes.

This preparation is composed of two operations: first, the rasping of potatoes; second, the separation of their fecula. Even sometimes, when the distiller does not choose to make his own fecula, he buys it from the feculist, who submits it to a third operation—that of draining.

These various operations will now be spoken of.

RASPING.

As a matter of course, the object of this operation is to tear the tissue of the vegetable, the cells of which contain the fecula which it produces, so that the better the vegetable is divided, the better the rasping will have been executed, and by these means a greater quantity of fecula will be obtained. To this effect, the potatoes are submitted to the action of a rasp, already described, and which, though it has not been brought to perfection, appears to be the best made use of.

Immediately after this operation, the following commences.

SEPARATION OF THE FECULA.

For this purpose, a large sieve of horse-hair, 2 feet in diameter and 1 foot in depth, is made use of. It is placed above a tub on two cross-pieces, and then filled with a quantity of pulp, equal to about two-thirds of its contents. This pulp is strongly rubbed between the hands, while a continuous stream of water, proportionate to the dimensions of the sieve, is running in the latter. For a sieve of 2 feet in diameter and 1 in depth, the water may be introduced through a pipe of 4 or 5 millimetres in diameter. This water, by means of the movement which the pulp undergoes, penetrates the latter, and runs through the sieve into the tub, carrying the fecula away in a state of dissolution.

This operation is continued until the water running through the sieve is clear and not impregnated with fecula. Then the pulp is thrown aside for the purpose of feeding cattle, and it is replaced by other, which is in the same manner deprived of its fecula. From 2500 kilogrammes of good potatoes 500 kilogrammes of fecula, supposed to be dry, are obtained, or 750 kilogrammes of drained fecula, which then bears the name of raw fecula. The latter is to the former : : 3 : 2, so that 3 kilogrammes of raw fecula will give 2 kilogrammes of dry; 13 hectolitres of pulp, or husks of potato, are moreover obtained, which contain about the same quantity of water as the raw potatoes—that is, three-fourths of their weight; so that those 13 hectolitres of pulp retain about 975 litres of water. This pulp may be given to cattle, but it is by far not so good as that resulting from the distillation of pastes, which is well boiled and nourishing.

It might be possible to obviate these inconveniences, in a distillery, by boiling the pulp with the hot spent-wash that is left in the still after distillation. There is a more suitable disposition of the sieve than that just indicated, and it is thought necessary to notice it here. It consists in filling at once with water the tub destined to receive the fecula; the sieve is suspended on the upper part of the tub with ropes, so as to be immersed in the water; a to-and-fro movement is given to it, by which means the pulp is freed from its fecula, which falls to the bottom of the tub. After this operation, the pulp is entirely exhausted by merely sprinkling a small quantity of water over it. This is but a modified disposition, which is thought useful, and which does not in the least alter the mode of working. The fecula, thus separated from the pulp, sinks, after an hour's rest, to the bottom of the tub, when the depth of the tub does not exceed 60 or 70 centimetres, and forms a thick and solid sediment. Holes made in the sides of the tub are then opened; that nearest the top of the sediment included.

The water runs off, and the fecula is left in the bottom If the latter is to be made in solid pieces, it is drained in the following manner.

DRAINING.

For this purpose, an apparatus called a *drainer* is made use of. It is a wooden chest, open on one side, widening in the shape of a cone toward the opening. The sides and bottom of this chest are filled with an infinity of small holes; the exterior part is covered with a cloth of strong linen or hemp. This drainer must be placed above the tub destined to receive the water that is to be drained; the fecula is then placed in the drainer, and in the course of twenty-four hours it loses all the superfluous water which it retained, and is found, it is true, in a state of humidity, but it is sufficiently solid to be transported without being made into paste. When the fecula is thought sufficiently drained, the drainer is turned over on a table used for this purpose, and there it is broken into pieces. This operation is only necessary, as already observed, when the fabrication of fecula is not connected with that of spirits.

This has been done at Paris, for instance, where many distillers buy their fecula from the starch-makers. From a given quantity of potatoes, 20 per cent. of dry fecula is obtained. No more can be reckoned upon, for the operation of rasping is not perfect enough to produce the result indicated in the chemical analysis.

PROCESS SPOKEN OF IN THE FIRST MODIFICATION.

For a tub of 12 hectolitres, intended to contain only 11 hectolitres of matter, from 80 to 85 kilogrammes of dry fecula, or from 120 to 125 kilogrammes of raw fecula, or all the fecula resulting from 400 kilogrammes or $5\frac{1}{3}$ hectolitres of good potatoes, are made use of. This quantity is deposited in the tub with a portion of cold water, so as to form a tolerably clear paste; that is, with about twice its weight of water. But great care should be taken to stir this mixture, because without this precaution the fecula, which is much heavier than water, would promptly precipitate itself to the bottom of the tub, and there form a hard sediment, which is with great difficulty brought to a state of suspension.

This state of suspension is necessary for a subsequent part of the operation. Every thing being thus disposed, and the necessary agitation having been given to the mass, from 500 to 600 litres of boiling water are gradually let into the tub; and before the whole of this hot water is in the latter, the mixture has already become thick, and has been converted into what is called paste. This paste has at first a milky appearance, but when the 600 litres of water are thoroughly mixed with it, the heat produced by them soon causes it to be remarkably clear and transparent. At this period the fecula is ready for being mashed, which is done by adding to it from 20 to 25 kilogrammes of malted barley or Indian corn, separately steeped.

In this instance the action of the malt is as conspicuous as in the mashing of rice; and in ten minutes, time requisite to mix the malt with the paste, the latter is completely fluidified. It is then left to subside for three or four hours, as in the mashing of corn, and after this time the liquor has acquired a very sweet taste. It is now sufficient to dilute it with water, so as to have 11 hectolitres of matter ready for fermentation. The advantage of this mode of working over that generally followed in this country is easily perceived.

The liquid, after having fermented, is found to be very fluid, and the sediment, which is composed of the husks of the corn and of a little portion of leaven, is very small. It may not be useless to remark, that in this case the malted barley should be ground into fine flour, and not merely broken, because its action on the fecula is thus more energetic, more perfect, and more rapid. The wash obtained by these means, and made with the given proportions of water and matter, marks about five areometrical degrees; 1 litre of good yeast is sufficient to bring it to fermentation.

PROCESS SPOKEN OF IN THE SECOND MODIFICATION.

The special object of this second modification is to avoid the labour occasioned by isolation of the fecula from the pulp. The following is the way of practising it with advantage and economy :—A double-bottomed tub, of about 8 hectolitres, is made use of. From 10 to 12 kilogrammes of chaff are spread on the first bottom, and the pulp, as it is produced from the raw potatoes, placed upon it; say, that obtained from 400 kilogrammes or $5\frac{1}{3}$ hectolitres of potatoes. There it is drained for half an hour; and thus a portion of water, naturally connected with its composition, is extracted without trouble. The latter is drawn off by means of the cock communicating with the space between the two bottoms. The mixture is then stirred, while from 400 to 500 kilogrammes of boiling water are gradually let in: the whole mass is now thickened : this change is caused by the conversion of the fecula into paste.

This mixture is then mashed with 25 kilogrammes of malt, previously steeped, and the liquid is left to subside for three or four hours. It is then drained and let into the fermenting-back, which contains 11 hectolitres. What is left is well drained for a quarter of an hour; then 2 hectolitres of boiling water are again let in. This mixture is agitated, drained, and taken with the rest to the fermenting-back. This lowers the temperature of the liquid. To cool and exhaust the paste completely, the whole surface of the sediment on the bottom of the tub is sprinkled with 1 or 2 hectolitres of cold water, which are drained and let into the fermenting-back with the other extracts. In operating thus, the husks left on the double bottom are sufficiently exhausted; they only retain, after an hour's draining, three-fourths of their weight in liquid, slightly charged with fermentable matter, which might be neglected in a well-conducted distillery, where the feeding of cattle is an object. But, at all events, if the loss of liquid was thought of any moment, the pulp might be entirely exhausted by means of a cylindrical press.

Full half of the weight of the drained pulp might be obtained in liquid, but a simple draining is sufficient, and the practice of pressing the residue should only be resorted to in cases where a smaller quantity of water has been used for mashing than has just been recommended. In this way of working, the fermentable matter being necessarily left to itself for a certain space of time, and often requiring to be taken from one tub to the other, is tolerably well cooled, and gains, toward the end of the operation, a suitable temperature for fermentation. There are, then, three ways of saccharifying the fecula of potatoes by means of malted barley or Indian corn.

It requires very little reflection to see that the first mode is evidently inferior to the other two. In fact, under whatever light it be considered, whether as a matter of economy in labour and combustibles, or with respect to the quality or quantity of the spirituous produce, it will never bear comparison to the other two. It is necessary that the potatoes should be boiled by steam, and this is indispensable. This steam is to be produced on purpose, and occasions expenses in fuel.

In the other two ways no combustibles are wanted to convert the fecula into paste proper to be mashed, and by working as heretofore advised for the mashing of grain; that is, to boil the matter intended for fermentation, whether it be isolated fecula or pulp, with the spent-wash left after distillation. In one case, the potatoes must be broken between two cylinders; in the two other cases, it is sufficient to rasp them; and this operation is not expensive, considering the great quantity that can be rasped by two men.

In the first case, the matter submitted to the still is heavy and pasty; it requires more combustibles to be brought to ebullition, and more expense is occasioned through the necessity of continually agitating it; and, without speaking of the bursting of the apparatus which it might cause, the spirit produced from it is always more or less charged with empyreuma. In the other cases, the matter is perfectly fluid, does not require any precaution to be taken in distilling it, and gives a better flavoured spirit. The superiority belongs, then, evidently, to the two last modifications that have been proposed; and, of those two, there can be no hesitation in declaring the latter to be the best. 100 kilogrammes of potatoes may thus produce from 50 to 55 litres of spirit at 75°. This spirit, as all potato spirit, has a small taste of the fruit, which is not pleasant, but of which it can be freed by *careful rectification*.

ARRACK, OR SPIRITS OF RICE.

Rice contains no sugar, and its composition appears to be essentially farinaceous. Carolina rice contains from 83 to 85 per cent. of its weight of fecula, or starch. To produce *arrack* from pure rice, it would be necessary to malt the latter, and, for this purpose, to submit it to all the operations connected with malting; that is, it should be steeped, malted, dried, and ground into flour. The only difference that would exist between this process and that of malting grain would be, that rice requires much more time to be sufficiently steeped and malted. As for the rest of the operations, they are exactly the same. Rice, being thus brought to the state of ground malt, may undergo a very good spirituous fermentation, whether it be mashed and fermented in the state of lob, or whether its farinaceous principles be extracted by means of the double-bottomed tub.

The distiller might adopt either of those two methods, according as he wishes to distil either heavy matter or such as is exempt from sediment. As to the method of fermenting rice prepared by either of those two processes, it is absolutely the same as in the fermentation of corn. The mashing requires the same temperature—the quantity of water made use of has the same influence over the spirituous produce; the only difference between the fermentable properties of rice and those of other corn is in the impossibility of causing it to ferment by solely preparing it by mashing. However, it might be sufficient only to malt a certain portion of the rice destined for distillation, and to mix it in the proportion of one-fourth or one-third of malted rice to three-fourths or two-thirds of unmalted; the fermentation would thus be equally complete. There is another method of predisposing rice to fermentation, which has been found successful. It is founded on the property which malted barley possesses of saccharifying the grain that is mashed with it.

Supposing that 80 kilogrammes of rice are to be worked, they are first reduced into fine and well-divided flour. This is thrown into a tub of about 12 hectolitres, and there it is diluted with 80 kilogrammes of water, such in temperature as to cause the thermometer, after the mixture has been well agitated and brought to a uniform mass, to rise to 77° or 80°. This mixture is left to subside for about half an hour, while 20 kilogrammes of malt, ground into fine flour, are separately steeped and well diluted in an equal weight of water at 100°. The mixture of rice having been left to itself during the time indicated above, the tub is uncovered, and boiling water is let into it until the mass becomes thicker and has the consistence of a dense lob; all this while the mixture is agitated until the thermometer has risen to 180°. Then the mashing commences; it is effected by throwing into the tub the portion of malt separately steeped.

The mixture is then agitated in all directions to render it homogeneous, and to establish a perfect contact between the malted barley and the rice. When this condition is evidently fulfilled, the tub is covered up again, and the wort is left to subside for three or four hours. At this period a phenomenon worthy of remark takes place: the mixture has undergone a total change, and the tub, which a few hours before contained a compact and pasty matter, now presents a liquid completely fluid, slightly sweet and saccharine; and there is only a small sediment formed at the bottom, which is composed of the husks of the grain and of the rice, and also of a few lumps, from one or the other of these vegetables, that have escaped decomposition. It is now sufficient to lengthen the liquid out with cold water, so as to bring it to 44° density, and to the temperature proper for fermentation.

This proceeds well through all its stages, and gives a vinous liquor, which is distilled without difficulty; the sediment being so small and of so little strength, that it may be neglected without any prejudice. This operation shows the utility of malt in the fermentation of corn in every light; and it is here evident that it possesses the property of converting the fecula, reduced into a paste by boiling water, into a kind of soluble matter which has all the properties belonging to sugar. This mode, then, is very practicable in the distillation of rice; it has the invaluable advantage of giving greater and better products, while it renders the wash perfectly fluid.

SPIRITS OF BEET-ROOT.

When we know that a vegetable body has in it saccharum, or sugar, we must take that as sufficient evidence that it possesses fermentable properties; and of course there is a possibility of drawing spirits from it. The sugar of the beet-root is identical with that of the cane when it is refined; consequently, it is quite as fine and as good, and does not cost the farmer much of an outlay. The production of solid sugar in the beet-root, as all other vegetable products, is subject to agricultural chances. Some years are more favourable to it than others; but an intelligent manufacturer, thoroughly acquainted with his art, will always escape great losses in a more or less fortunate way.

So it is, for instance, that a manufacturer of beet-root sugar, finding in unfavourable years that the small quantity of sugar which the vegetable gives him would not defray his expenses of fabrication, meets with a precious resource in submitting it to distillation. The choice of the beet-root, either to make sugar or produce spirit, is not more indifferent in one case than the other. There exists a great variety of them, all of which are distinguished by the colour of their peel and that of their pulp.

The white, the yellow beet-root, and that which is white inside and red outside, are preferable to all others. Whatever be the colour of the root, it is essential to appropriate it to the soil, to cultivate it in a fit and proper manner, and not to use the seeds of degenerate species Experience has proved the *streaky beet-root* to be the least productive, and it must of course be rejected as much as possible. The beet-root thrives in mixed soils: by this is meant such as are not too mobile nor too argillaceous, neither too calcareous nor too sandy.

The soil should not be too damp nor too dry. It grows well after all sorts of manures, sparingly distributed; however, strawy manures and the stalling of herds are more proper for it. It is generally sown in the month of April, and gathered toward the end of September or beginning of October, according to the climate. As soon as the beet-roots are drawn out of the ground, they are necked and put up where the frost cannot injure them, in cellars or in pits dug four or five feet deep, covered afterward with a layer of earth from one to one and a half foot in thickness. Then they are taken as wanted, and their juice is expressed by means of a *rasp*, which will now be described; this being, therefore, the most suitable place, we will now speak of

THE BEET-RASP.

This is made as follows:—A frame made of oak, built strongly, having an oblong form, mounted on four legs joined together from top to bottom by cross-pieces, constitute the assemblage bearing the various parts of the mechanism; nearly all of them disposed on the length of the upper cross-pieces. Those parts are composed of a wooden cylinder, made plain and suitably prepared. Its diameter is 18 inches, and its breadth 8 inches; its circumference is armed with 80 saw-blades, 7 inches long

On one of the extremities of the axis of the cylinder is an iron pinion, furnished with 16 teeth, working those of a wheel, also of iron, and having 120 teeth; a handle of 18 inches is fixed on each extremity of the axis of this wheel. Under this cylinder is placed a kind of tank, inclined in such a manner as to send the pulp obtained into a bucket filling the functions of a recipient; on the same face of the frame, and before the circumference of the cylinder, is adjusted on a mobile centre a kind of wooden shutter, which receives, from the axis of the pinion and by the aid of swing-gates, a to-and-fro motion, in such a manner as alternately to open and close the interval left between the cylinder and that same shutter for the passage of the beets or potatoes.

Nevertheless, the opening is limited by means of a little bar, on which the shutter rests in its back motion. All the parts of this machine, exceeding the frame, are enveloped in a box, surmounted by a hopper yielding about a quintal of potatoes or beets. From this kind of cage it results that the trituration is effected, very properly, without splashing or loss of matter. 2500 kilogrammes* of beets or potatoes may be reduced in twelve hours by this rasp, worked by only two men.

The pulp is then placed in bags, which are to be pillared and submitted to the action of a screwing or hydraulical press: this kind of press is preferable to any other, because it possesses the greatest force. By these means a quantity of the juice is obtained proportionate to the energy of the machine made use of. With a cylindrical

---

* A kilogramme is about two pounds.

press, well mounted and well conducted, it is possible to obtain a quantity of juice equivalent to 75 per cent. of the raw beet-root. To effect this it is necessary to wet the residue of the first extract, and to submit it again to the action of the press. With screwing or hydraulic presses, 65, 70, and even 80 per cent. of the juice can be obtained in one operation, according to the power of the engine, or the more or less aqueous qualities of the roots. This juice, supposing no water to have been used, may mark from 5° to 9° Beaumé, according to years and the species of fruit.

It contains, like the cane, two kinds of sugar—one solid, and the other liquid; that is, one that can be crystallized, and the other molasses. It contains, besides, water, leaven, and some extractive substances, one of which gives to the beet-root that acrid taste by which it is distinguished, and also the property of attacking the throats of those who eat it; this is not entirely removed even by the operation of boiling, as many persons can testify. This substance might communicate to the spirits of beet-roots its noxious taste, was it not corrected by the operation of rectification, which has already been spoken of. The liquid separated by means of the press may be put immediately to fermentation; leaven enters into its composition, and soon begins to work. A good soil may produce from 80,000 to 100,000 pounds of beet-roots per hectare.

The molasses of beet-roots, as before observed, has not been applied to any other use than that of distillation. This results from its peculiar bad taste, which is the cause of its being rejected by the trade. In fact, it cannot be delivered for consumption in the state it is in, like the molasses from the sugar-houses; there is nothing of which the former partakes less than of that luscious savour of the latter; and this is the only difference existing between the raw sugar of the beet-root and that of the cane, both of which are identical after being purified of their molasses.

The molasses of beet-roots takes with it all the acridity of the root; and, moreover, it happens very often to have a strong taste of salt, caused by the nitrate of potash, or saltpetre, which the beet-root contains in large quantities. The molasses of beet-roots produces more spirit than the sugar-cane molasses. Its taste, it is true, resembles less that of rum, and always keeps a particular aroma; but it is one agreeable to the palate, and might, perhaps, with certain precautions, be rendered more identical with that of the rum made from sugar-cane molasses.

The method made use of for distilling this spirit is conducted as follows; this is a method followed in preparing beet-root molasses for a good fermentation :—100 litres* of molasses are mixed with 100 litres of boiling water. When all has been well blended, the back is covered, and the mixture left to itself for twelve hours. Then 2 hectolitres of boiling water are added, the mass mixed well, and left to repose for twelve hours more. At this period proceed to the fermentation; and, to effect this, dilute the whole mass with warm and cold water, so as to bring it to the temperature required, and to give it from 5 to 6 areometrical degrees density.

---

* A litre is about two wine-pints.

From 1 hectolitre of syrup you can obtain 20 or 24 hectolitres of well-fermented wash, which renders about 80 litres of spirits at 19° Beaumé. This quantity differs from those reported in various works—some saying more could be produced, and others, that not so much might be looked for under any circumstances. The medium has been taken here, which is more likely to be right than either of the extremes alluded to.

The only way to account for the results so widely differing is by the admission of the fact, which is very probable, that by exhausting the molasses much more, it is rendered less fit for distillation; while another operates on molasses *richer in sugar*, less exhausted, and with a better knowledge of that particular kind of work. It is necessary to observe that the produce of spirits mentioned before is owing to the process which has just been described for fermenting it. Moreover, the reader must be warned that one of the advantages attached to this method of operating results from the use of the spent-wash.

This occasions an economy in fuel, as the caloric of the wash, leaving the still in a boiling state, is in this instance appropriated to good use. Besides, there are found in the use of the spent-wash all the benefits which have been before developed in speaking of the transformation of sugar into alcohol.

It has often been found of advantage to put grain to this mixture, at the rate of from 5 to 7 kilogrammes per 100 litres of molasses. This grain, when broken and consisting of a mixture of 20 kilogrammes of malt to 80 of rye, gives more impetus to the fermentation, and renders it more complete.

KIRSCH-WASSER, OR THE SPIRITS OF CHERRIES.

There is a kind of spirits, prepared from cherries, known by the name of *kirsch-wasser*. The name comes from the German, and is composed of two words signifying "water of cherries." This liquor is made to the greatest extent in Switzerland and Germany, where cherry-trees are very common; that consumed in France comes from the neighbourhood of the *Forêt Noire*. The way of preparing the worts is as follows:—In the month of July or August, when cherries have arrived at maturity, no trouble is taken to pick them, and they are brought down by means of poles, which is decidedly a bad practice, because it damages the trees; and the cherries, leaves, and small branches all fall down together, which, gathered by children, are deposited in a trough, without any precaution, so that the spoiled and good ones are all mixed together. This trough represents a cylinder more wide than deep, and yielding according to the extent of the fabrication.

It is placed above the fermenting-back on two girders, which rest on the borders of the back, and are joined together by two cross-pieces of the same thickness. This trough being half or three-quarters full, men, women, and even children place themselves around it, and proceed to the pressing of the cherries with their hands, so as to set the juice at liberty. This cannot be done perfectly, as it may well be supposed, by squeezing the cherries between their hands, or by rubbing them against the sides of the trough.

The juice runs then into the fermenting-back through the interstices of the trough, while skins, stalks, and stones are left behind. The stones are then added to the liquid, and the wort is left to ferment. It was thought for a long time that it was necessary to break the stones, from which the "*kirsch*" derives its characteristic flavour and aroma, to obtain this effect; but experience has, in a conclusive manner, demonstrated that this practice is useless, and that the worts from cherries fermented with the stones, either entire or broken, give an equally good-flavoured kirsch.

Kirsch being only consumed as a liquor, the fabrication of it is not very extensive, and the wine producing it is put to the still immediately after the fermentation, which lasts from six to eight days. Leaven is put with this wort.

In some parts of the United States there are immense numbers of what are called "wild-cherry trees," which bear a large quantity of fruit. There can be no doubt but they could be made useful in the same way as other cherries; and, from my knowledge of them, I think a fine spirit could be made from the fruit by the process just spoken of. At least, it is well worthy of a trial.

OF SOME OF THE PRODUCTS OF THIS COUNTRY WHICH AFFORD SPIRITS BY DISTILLATION.

This country abounds in many fruits, roots, and vegetables that will yield spirit upon distillation. It would be useless to give a separate process for every one of those substances, as similar substances require a similar mode of treatment. This, it is hoped, will be a sufficient hint to any one who may wish to experiment on a particular kind of fruit, vegetable, &c. And first will be spoken of—

CIDER SPIRITS, OR APPLE BRANDY.

In many parts of the United States large quantities of apples are raised, which cannot be made use of to advantage in any other way, and it therefore becomes an object to the farmer to distil them. The process is worked thus:—The apples, after being assorted, so as to work the ripest first, are then ground, either in the common way, or with a mill constructed similar to the tanner's bark-mill; after which they are pressed in a large, powerful screw-press as long as any juice can be obtained.

The cider is then put into large cisterns or vats prepared for the purpose, where it undergoes a fermentation, and is fit for the still in from six to twelve days, according to the weather. Some distillers preserve the pomace of the pressing, put it into casks, and cover it with water, until it undergoes a fermentation, when it is again pressed out, and the cider distilled. This, however, requires so much work and so many casks, that in a busy season it is scarcely worth attending to; but when fruit is scarce, it may be done.

Many persons are in the habit of grinding the apples, and then throwing them into casks, where they undergo a fermentation, after which the whole mass is committed to the still. Though a greater quantity is said to be obtained in this way than any other, it is a bad plan, as the brandy is certain to possess that peculiar empyreumatic taste which renders it very unpalatable. The operation is also more tedious, and, upon the whole, the least profitable.

To judge of the progress of fermentation, run a stick down in the centre of the cask; if, upon drawing it out, it is accompanied with a bubbling, hissing noise, the fermentation is not over; but if no such noise is observable, it is then fit for the still. To those who are desirous of following this plan, it is advised, as the best method of avoiding an empyreuma, that the still be one-third filled with water, which must be made to boil before putting in the pomace. The spirit made from cider is in every respect better than that made from pomace.

PEACH BRANDY

Peaches grow in great abundance in nearly every part of the United States, but more abundantly and of a better quality in the Southern States. The flavour of peaches is equal, if not superior, to that of any fruit in the world.

Upon distillation they yield a spirit remarkably fine and agreeable, which is made use of very much in the mixing of liquors. The methods of treating peaches and apples are similar. By some, the fruit is thrown into a large trough, where it is pounded with large pestles until completely mashed; it is then pressed out; and a hogshead of pure juice, obtained in this way, will yield from 10 to 12 gallons of the best brandy. As the pomace cannot be completely pressed, it is thrown into casks, diluted with water, and, after sufficient fermentation, again pressed, and immediately distilled.

Another method, and the best, where a large quantity of peaches are distilled, is to grind them in a suitable mill, which, by mashing the stone and kernel, is said to impart an agreeable bitter to the spirit. In this state it is fermented, and, with the addition of a small quantity of water, committed to the still. Others press it after the manner of pressing apples, which is far preferable to all other modes.

OF THE PREPARATION AND DISTILLATION OF RUM.

It is necessary to remark, in the beginning, that in the still-house, as well as the boiling-house, the greatest cleanliness is requisite. The vats, at the beginning of the crop, ought to be well washed out, with both warm and cold water, to divest them of any sour stuff which may have accumulated or adhered to their bottoms and sides since they were last in use; and if every vat, just before the first setting, or mixing the liquor in it, were to be rinsed with a little rum, the distiller would be well repaid for this small outlay and trouble. In sitting the first round of liquor, a greater proportion of skimming from the sugar-pans must be used than will afterward be necessary, as the distiller has no good lees, and very little molasses, to add to the mass; and, besides, the skimmings at this time are not so rich as they will be some time hence—in March, April, or May, which are thought the best yielding months.

The following proportions will succeed well in the beginning:—For every 100 gallons your vat contains, put 45 gallons of skimmings, and 5 gallons of molasses to 50 gallons of water. When you have got good lees, or returns, as they are often called, mix equal quantities of skimmings, lees, and water, and for every 100 gallons add 10 gallons of molasses. When the mill is going, and therefore you have no skimmings, mix equal parts of lees and water, and for every 100 gallons add 20 gallons of molasses. From liquor set in these proportions the distiller may expect to obtain from 10 to 15 per cent. of proof-rum, and twice as much low wines.

But the quantity of spirit will depend greatly on the quality of the ingredients, and in some measure on the weather; therefore, an intelligent distiller will vary his proportions accordingly. Rum differs from what is simply called sugar-spirit, as it contains more of the natural flavour or essential oil of the sugar-cane; a great deal of raw juice, and even parts of the cane itself, being often fermented in the liquor or solution of which the rum is prepared. For this reason it is generally thought that the rum derives its flavour from the cane itself. Some, indeed, are of opinion that the oily flavour of the rum proceeds from the large quantity of fat used in boiling the sugar. This fat, of course, will give a rancid flavour to the spirit in distillations of the sugar-liquors, or wash, from the refining sugar-houses; but this is nothing like the flavour of rum.

Great quantities of rum are made at Jamaica, and other places in or near the same latitude; the method of making it is this:—When a sufficient stock of materials is got together, they add water to them, and ferment them in the common way, though the fermentation is always carried on very slowly at first, because, at the beginning of the season for making rum in the islands, they want yeast to make it work; but after this they, by degrees, procure a sufficient quantity of the ferment, which rises up as a head to the liquor in the operation; and thus they are able afterward to ferment and make their rum with a great deal of expedition, and in very large quantities. When the wash is fully fermented, or to a due degree of acidity, the distillation is carried on in the common way, and the spirit is made up proof, though sometimes it is reduced to a much greater degree of strength, nearly approaching to that of alcohol, or spirits of wine; and it is then called "double-distilled" rum.

There can be no doubt that it would be easy to rectify the spirit, and bring it to a much greater degree of purity than it is usual to find it, if it did not bring over in the distillation so large a quantity of the gross oil, which is often so disagreeable that the rum must be suffered to lie by a long time to mellow before it can be used; whereas, if well rectified, its flavour would be much less rancid, and consequently much more agreeable to the palate.

It has been ascertained that the best state to keep rum, both for exportation and other uses, is doubtless in that of alcohol, or rectified spirits. In this manner it would be contained in half the bulk it usually is, and might be let down to the common proof strength with water, when necessary.

PROCESS MADE USE OF IN GREAT BRITAIN AND IRELAND FOR FERMENTING AND DISTILLING MOLASSES.

THIS process will be found well adapted to the use of those of our citizens who are not living in the sugar-growing regions. It is conducted as follows:—They set the backs in the former (Great Britain) by adding 2 gallons of water and 1 of molasses; to which (in both places) they add about 1 gallon of barm or yeast to 200, and sometimes 300, of molasses so mixed. These they blend, with a large birch-broom, uniformly together; this they call setting.

This must be attended to once or twice a day, and the head stirred in or more barm added occasionally; or the air partially excluded to keep it warm, if it works slow, and admitted fully, if it works fast. In three or four days the backs must be raised, by adding (in Great Britain) 2

DISTILLING MOLASSES.

gallons of water more to each gallon of molasses set; and in Ireland the same; consequently, they work their wash one-fifth stronger in Great Britain than in Ireland: and when they wish to evade the duty of excise, they work their wash still stronger, but this materially hurts the quality of the produce.

In the winter time, the water added to the backs should be heated to a degree below blood warm, that the backs are raised with, which may be done by heating some water scalding hot, not boiling it, in one of the stills, and drawing as much in the filling-can as will heat the remainder of the cold water to the degree wanted. When the intended portion of water is added to each back, the same proportion of barm is to be added as at setting, and all blended together with the broom; this is termed raising.

The same, or rather more, attention must be paid after setting, and barm added, if necessary. The third stage of fermentation is cutting, which is performed four, five, or even six days after raising, but is seldom deferred so long. It is done by adding about 1 ounce of jalap-root, in fine powder, to every 800 or 1000 weight of molasses in summer, and half as much more to the same quantities in winter, with the same proportion of barm, or yeast, as at setting and raising, which must be blended together with the yeast. This is called cutting the backs, which, indeed, it very effectually does—cutting down the head or crest of the flowers or barm which the intestine motion of the fermentation threw up, and communicating a very effectual ferment-essence through the whole fluid mass, very distinguishable at the top of the fluid to the sight, and also to the ear; the hissing of which can now be distinctly heard by those who are near.

As this tumultuous motion and hissing noise lessens, the operation draws to a close; and when they can be no longer distinguished, which is generally in three or four days after cutting, the fermentation is over, and the fermented wash is to be emptied into the still, and the backs set anew, as before directed. This fermented wash, distilled as long as a glass of it, thrown upon the still-head, will burn or take fire from a lighted paper or candle, is called low wines, or spirits of the first extraction. These low wines are kept for three distillations, which quantity generally fills the still, which is called doubling, or second extraction, and are drawn off as directed previously.

This spirit, lowered with water to the hydrometer standard, is called proof-spirit. After the setting of the backs, if an addition of barm does not bring on a sensible fermentation through the whole, a five-gallon can of warm spent-wash, added to every 200 gallons of the fermenting-wash, will in general bring on the desired degree of fermentation; if not, about half the quantity of jalap usually used in cutting the backs must be added now, and the other half at cutting the backs. In winter, particularly in frosty weather, the part of the still-house where the fermentation is going on must be heated to the temperature of temperate on the thermometer, which will much facilitate the process. This may be done by the heat of the stills at work in winter; and the excess of heat from them in summer may be counterbalanced by windows contrived to draw a current of air across the still-house.

RAISIN SPIRITS.

From raisins is extracted a spirit, after proper fermentation, bearing this name. In order to extract this spirit, the raisins must be infused in a proper quantity of water and fermented. When the fermentation is completed, the whole is to be thrown into the still, and spirits extracted by a strong fire. The reason why a *strong* fire is here directed is, because by that means a greater quantity of the essential oil will come over the helm with the spirit, which will render it much fitter for the distiller's purpose; for this spirit is generally used to mix with common malt goods; and it is surprising how far it will go in this respect, 10 gallons of it being often sufficient to give a determining flavour and agreeable vinosity to a whole piece of malt spirit.

FLAVOURING AND COLOURING OF SPIRITS.

The sweet spirits of nitre, either strong or dulcified, is the substance generally used by distillers for the flavouring of spirits, to deprive them of their lixivious taste after rectification. As regards the *colouring* of spirits, that of French brandy has been held up as the acme of perfection. The extract of oak has been proposed; but after all, the most practical means found by experience is the use of common treacle and burnt sugar, though it has been said that neither of these will succeed when put to the test of the vitriolic solution.

A quantity of oak-bark shavings, deposited for some time in spirits of wine, will form a dilute tincture of oak; this may be added to colour spirits, instead of burnt sugar. 1 pint of parched or burnt wheat will give an agreeable colour to 1 barrel of whisky, and will improve the flavour.

PROCESS FOR MAKING RUM SHRUB.

To effect this, take 65 or 70 gallons of rum, from 7 to 8 gallons of lemon-juice, 6 or 7 gallons of orange-juice, (both fresh expressed from the fruit,) orange-wine 30 gallons, 2 pounds of the rind of fresh lemon-peel, and 1 pound of the rind of fresh orange-peel, (both pared off as thin as possible, and previously steeped for a few days in the rum,) and 100 pounds of loaf-sugar. Fill up the cask, of 120 or 130 gallons, with pure spring-water; rouse them well together. If not sweet enough, sweeten to suit you; if too sweet, add more lemon-juice.

Dissolve your sugar in part of the water used for making up your shrub; let it stand till fine, set up on end, with a cock near the bottom.

PROCESS FOR MAKING BRANDY SHRUB.

This is done in the following way:—Take from 75 to 80 gallons of brandy, 8 or 10 gallons of lemon-juice, 8 gallons of orange-juice, 4 pounds of thin rind of fresh lemon-peel, and 2 pounds of orange-peel, fresh, (both pared as thin as may be,) and add them to the brandy the first thing; with 4 ounces of terra-japonica, 1 hundred-weight of loaf-sugar or clayed sugar, dissolved in part of the water used for making up, added with the above ingredients to the brandy, &c. Fill up with good clear water, set the cask on end, with a cock near the bottom, and let it stand till fine.

Shrub may be made in a similar manner with whisky, apple brandy, peach brandy, &c., with similar ingredients in the before-mentioned proportions. The quantity can be increased or reduced to suit the operator, by duly proportioning the ingredients to the quantity of spirits employed.

ELDER JUICE.

To make this article, you must let your berries be fully ripe, and all the stalks (which are numerous) be clean picked from them. Then, if you have a press for drawing all the juice from them, have ready four hair-cloths somewhat broader than the press, and put one layer above another, having a hair-cloth between every layer, which must be laid very thin and pressed, first a little, then more, till your press be drawn as close as you can get it; then take out the berries, and press all you have in like manner. Then take the pressed berries, and break out all the lumps; put them into an open vessel, and put on them as much liquid as will just cover them Let them infuse so for seven or eight days; then press it out, and either add to it the rest, or keep it separately for present use, and put your best juice into a cask proper for it to be kept in; and put 1 gallon of malt spirits, not rectified, to every 20 gallons of elder juice, which will effectually preserve it from becoming sour for two or three years.

METHOD OF MAKING CHERRY BRANDY.

There are several ways of making this liquor, which is in great demand. Some press out the juice of the cherries, and having dulcified it with sugar, add as much spirit to it as the goods will bear, or the price it is intended to be sold for. But the common method is to put the cherries, clean picked, into a cask with a proper quantity of proof-spirit; and after standing about eighteen days, the goods are drawn off into another cask for sale, and two-thirds of the first quantity of spirits poured into the cask upon the cherries. This is to stand one month, to extract the whole virtue from the cherries; after which it is drawn off as before, and the cherries pressed, to take out the spirit they had absorbed.

The proportion of cherries and spirit is not very nicely observed; the general rule is that the cask be half-filled with cherries, and then fill up with proof-spirits.

Some add to every 20 gallons of spirit half an ounce of cinnamon, 1 ounce of cloves, and about 3 pounds of sugar, by which the flavour of the goods is considerably increased. But, in order to save expenses, not only the spices and sugar are generally omitted, but also a great part of the cherries, and the deficiency supplied by the juice of elder-berries. Your own reason, therefore, and your taste, or the price you intend to ask for it, must direct you in the selection of your ingredients.

By the same method you can make raspberry brandy · should the colour of the article not be so deep as you wish, it can be made more so by the addition of a little cherry brandy, elder juice, or other colouring substance, such as logwood, &c.

EAU DE LUCE.

THE process for making this is simple and easy of execution. Take of the oil of amber 1 ounce, of highly-rectified spirits of wine 4 pounds; put them into a bottle, and let them remain there five days, shaking the bottle occasionally during the time, by which means the spirit will be strongly impregnated with the oil. Then put into this impregnated spirit 4 ounces of choice amber, finely powdered, and let it digest three days; thus you have a very rich tincture of amber. The tincture being thus made, take of the strongest spirits of sal-ammoniac 16 pounds, and add to the foregoing tincture, together with 8 pounds of highly rectified spirits of wine. You will thus obtain the celebrated "*Eau de Luce,*" which is so much in use in all cases of fainting, lowness of spirits, giddiness, headache, &c.

IRISH USQUEBAUGH.

This is a very celebrated cordial, the basis of which is saffron. Take of nutmegs, cloves, and cinnamon, of each 2 ounces; of the seeds of anise, caraway, and coriander, each 4 ounces; liquorice-root, sliced, half a pound. Bruise the seeds and spices, and put them, together with the liquorice, into the still, with 11 gallons of proof-spirit and 2 gallons of water; distil with a pretty brisk fire till the feints begin to rise. But as soon as your still begins to work, fasten to the nose of the worm 2 ounces of English saffron, tied up in a cloth, that the liquor may run through it and extract all its tincture; and in order to do this, you should often press the saffron with your fingers. When the operation is finished, dulcify the spirits with fine sugar.

This may be prepared without distillation in the following manner:—Take of raisins, stoned, 5 pounds; figs, sliced, 1½ pound; cinnamon, half a pound; nutmegs, 3 ounces; cloves and mace, of each 1½ ounce; liquorice, 2 pounds; saffron, 4 ounces. Bruise the spices, slice the liquorice, and pull the saffron in pieces; digest these ingredients eight days in 10 gallons of proof-spirit, in a vessel close stopped. Then filter the liquor, and add to it 2 gallons of canary wine and half an ounce of the tincture of verdigris.

PROCESS OF MAKING NECTAR.

This may be made with 15 gallons of the "imperial ratafia," a quarter of an ounce of cassia oil, and an equal quantity of the oil of caraway seeds, dissolved in half a pint of spirits of wine, and made up with orange wine, so as to fill up the cask. This process is for making 20 gallons. Sweeten, if wanted, by adding a small lump of sugar in the glass.

IMPERIAL RATAFIA.

Take three-quarters of a pound of the kernels of peaches, nectarines, and apricots, bruised; 3 pounds of bitter almonds, bruised; half a gallon of rectified spirits of wine, in which dissolve half an ounce of compound essence of ambergris; 12 gallons of pure molasses spirit, and as many gallons of rose-water as will make up the ratafia to 20 gallons. Steep the kernels and almonds for ten days; then draw off for use.

This quantity will take 10 pounds of loaf-sugar to sweeten it; but as some may not like it so, it had better be sweetened by a few gallons at a time, as it is wanted.

METHOD OF MAKING LOVAGE CORDIAL.

This cordial, which has been in use for a long time, can be made thus:—Take of the fresh roots of lovage, valerian, celery, and sweet-fennel, each 4 ounces; of essential oil of caraway and savin, each 1 ounce; spirits of wine, 1 pint; 12 gallons of proof-spirits; loaf-sugar, 12 pounds. Steep the roots and seed in the spirits fourteen days. Dissolve the oils in the spirits of wine, and add them to the undulcified spirit cordial drawn off from the other ingredients; dissolve the sugar in the water for making up; fine, if necessary, with alum.

PROCESS OF MAKING CITRON CORDIAL.

Take of Smyrna figs, 14 pounds; spirits, 12 gallons. Infuse for one week; draw off, and add to the clear spirituous infusion essence of orange and lemon, each 1 ounce, dissolved in a pint of spirits of wine; half a pound of dried lemon, and 4 ounces of orange-peel; 6 or 7 pounds of loaf-sugar. Make up, as before, with clean, nice water.

CINNAMON CORDIAL

This very agreeable compound is useful in families, being often sufficient to arrest sickness at the stomach, &c. &c. It is thus made:—Take 1 drachm of oil of cassia, dissolved with sugar and spirits of wine; 1½ gallon of spirits; cardamom-seed, husked, 1 ounce; orange and lemon-peel, dried, of each 1 ounce. Fine with half a pint of alum-water; sweeten to your taste with loaf-sugar, not exceeding 2 pounds, and make up 2 gallons measure with the water you dissolve the sugar in. This cordial can be coloured, if desired, with burnt-sugar.

FRENCH NOYAU.

Take of fine French brandy 1½ gallon; 6 ounces of the best fresh prunes; 2 ounces of celery; 3 ounces of the kernels of apricots, nectarines, and peaches, and 1 ounce of bitter almonds, all gently bruised; essence of orange-peel and lemon-peel, of each half a drachm, dissolved in spirits of wine; half a pound of loaf-sugar. Let the whole stand fourteen days; then draw off, and add to the clear noyau as much rose-water as will make it up to 2 gallons, which will be near half a gallon.

PEPPERMINT CORDIAL.

As this is easily and cheaply made, every family should make it for their own consumption. Take of rectified spirits 13 gallons; 12 pounds of loaf-sugar; 1 pint of spirits of wine; 15 pennyweights (troy) of oil of peppermint; water, as much as will fill up the cask, (20 gallons;) which should be set up on end after the whole has been well roused, and a cock for drawing off placed in it.

PROCESS OF MAKING ANISEED CORDIAL.

Take of spirits 14 gallons; spirits of wine, 1 pint; from 6 to 8 pounds of loaf-sugar; 1½ ounce of oil of aniseed; 2 ounces of finely powdered alum. Dissolve the sugar in one part of the water used for making up, and the alum in the remainder, and proceed as directed in the making up of peppermint cordial. Aniseed cordial does not bear to be reduced below one in five, as part of the oil will separate when too much lowered, and render the goods quite unsightly indeed.

METHOD OF MAKING CARAWAY CORDIAL.

This is done by taking of oil of caraway 1 ounce; oil of cassia, 20 drops; essence of orange-peel, 5 drops, and the same quantity of essence of lemon; 13 gallons of spirit; 8 pounds of loaf-sugar. Make it up and fine down as directed for aniseed cordial.

FRENCH VINEGAR.

Wine which is detained for this purpose is mixed in a large tun with a quantity of wine-lees, and the whole being transferred into cloth sacks placed within a large iron-bound vat, the liquid matter is extended through the sacks by superincumbent pressure. What passes through is put into large casks set upright, having a small aperture at their tops. In these it is exposed to the heat of the sun in summer, or to that of a stove in winter.

Fermentation comes on in a few days. If the heat should then rise too high, it is lowered by cool air and the addition of fresh wine. The art of making good wine-vinegar consists in the skilful regulation of the fermentative temperature. In summer, the process is generally completed in a fortnight; in winter, double the time is requisite. The vinegar is then run off into barrels containing several chips of birch-wood. It is clarified in about two weeks; and, to be fit for the market, must be kept in close casks.

MODE OF MAKING ENGLISH VINEGAR.

This is generally made from malt. By mashing with water, 100 gallons of wort are extracted, in less than two hours, from 1 bushel of malt. When the liquor has fallen to the temperature of 75° Fahr., 4 gallons of the yeast of beer are added. After thirty-six hours it is racked off into casks, which are laid on their sides, and exposed, with their bung-holes loosely covered, to the sun in summer, but in winter they are arranged in a stove-room.

In three months this vinegar is ready for the manufacture of the sugar of lead. To make vinegar for domestic use, however, the process is somewhat different. The above liquor is racked off into casks placed upright, having a false cover, pierced with holes, fixed at about a foot from their bottom. On this a considerable quantity of *rope*, or the refuse from the makers of British wine, or, otherwise, a quantity of low-priced raisins, is laid. The liquor is turned into another barrel every twenty-four hours, in which time it has begun to grow warm. Sometimes, indeed, the vinegar is fully fermented as above, without the *rope*, which is added at the end to communicate the flavour. Good vinegar can be made from a weak syrup of 18 ounces of sugar to every gallon of water; yeast and rye are to be used as above described. Vinegar obtained by the preceding methods has more or less of a brown colour, and a peculiar but rather grateful smell.

SOME GENERAL DIRECTIONS FOR THE DISTILLATION OF SIMPLE WATERS.

It must constantly be borne in mind that plants and the parts which are to be used ought to be fresh gathered. Where they are directed fresh, such only must be employed; but some are allowed to be used dry, as in this state they may easily be procured at all times of the year, though more elegant waters might be obtained from them while quite green. Having bruised the substances a little, pour thereon thrice their quantity of spring-water. The quantity, however, may be diminished or added to, according as the plants may be more or less juicy than ordinary. When fresh and juicy herbs are to be distilled, thrice their weight of water will be quite sufficient, but dry ones require a much greater quantity. In general, there should be so much water that, after all intended to be distilled has come over, there may be liquor enough left to prevent the matter from burning to the still. Formerly, some vegetables were slightly fermented with the ordinary yeast previous to distillation. Should any drops of oil swim on the surface of the water, they are carefully skimmed off. That the waters may be kept the better, about one-twentieth part of their weight of proof-spirit may be added to each after they are distilled.

Such is a short but accurate and complete sketch of the distillation of simple waters.

OF THE STILLS USED FOR SIMPLE WATERS.

There are not a great many instruments used for this purpose; those chiefly in use are of two kinds—commonly called the hot still, or alembic, and the cold still. The waters drawn from plants by the cold still are much more fragrant, and more fully impregnated with their virtues, than those drawn by the hot still, or alembic. The method is this:—A pewter body is suspended in the body of the alembic, and the head of the still fitted to the pewter body; into this body the ingredients to be distilled are put, the alembic filled with water, and the still-head luted to the worm of the refrigerator. The same object would be fulfilled by putting the ingredients into a glass alembic, and placing it in a bath heat, or *balneum mariæ*. The cold still is much the best adapted to draw off the virtues of simples which are valued for their fine flavour when green, which is subject to be lost in drying; for when you want to extract from plants a spirit so light and volatile as not to subsist in open air any longer than while the plant continues in its growth, it is certainly the best method to remove the plant from its native soil into some proper instrument, where, as it dies, these volatile parts can be collected and preserved.

Such an instrument is what is called the cold still, where the drying of the plant or flower is only forwarded by a moderate warmth, and all that rises is collected and preserved. As the method of performing the operation by the cold still is the very same, whatever plant or flower is used, the following instance of procuring a water from rosemary will be sufficient to instruct the young practitioner in the manner of conducting the process in all cases whatever :—Take of rosemary, fresh gathered in its perfection, with the morning dew on it, and lay it slightly and unbruised upon the plate or bottom of the still; cover the plate with its conical head, and apply a glass receiver to the nose of it.

Make a small fire of charcoal under the plate, continuing it as long as any liquor comes over into the receiver. When nothing more comes over, take off the still-head and remove the plant, putting fresh in its stead, and proceed as before; continue to repeat the operation successively till a sufficient quantity of water is procured. Let this distilled water be kept at rest in clean bottles, close stopped, for some days, in a cold place. By this means it will become limpid and powerfully impregnated with the taste and smell of the plant. In this water is contained the liquor of dew, consisting of its own proper parts, which are not without difficulty separated from the plant, and cleave to it even in drying. This dew also, by sticking to the outside, receives the liquid parts of the plant, which, being elaborated the day before, and exhaled in the night, are hereby detained, so that they concrete together into one external liquid, which is often viscid, as appears in manna, honey, &c.

CINNAMON WATER.

Take of cinnamon 1 pound; water, 1½ gallon. Steep them together for two days, and then distil off the water till it ceases to run milky.

PEPPERMINT WATER.

Take of peppermint leaves, dry, 1½ pound; water, as much as will prevent the leaves from burning. Draw off by distillation 1 gallon.

DAMASK-ROSE WATER.

Take of damask-roses, fresh gathered, 6 pounds; water, sufficient to prevent the roses from burning. Distil off 1 gallon of the water.

ORANGE-FLOWER WATER.

Take 2 pounds of orange-flowers, and 24 quarts of water. Draw over 3 pints.

ORANGE WINE.

This delightful beverage is prepared in the following manner:—Take 12 oranges, and pare them very thin; strain the juice, so that none of the seeds go in with it. Then take 6 pounds of loaf-sugar, and the whites of 2 eggs, well beaten; put these into 3 gallons of spring-water, and let it gently boil for half an hour. As the scum rises, take it off; then add the orange-juice and rind.

Three or four spoonfuls of yeast must also be put in, and let it stand in a pan or pail for four or five days; then put it into the cask, and let it stand for three or four weeks, but do not stop it close for the first week. When nearly fine, draw it off into another cask, and add to it a quart of white wine and a little Cognac brandy. Stop it close, and in a month or six weeks it will be in fine condition, ready for use.

SIMPLE LAVENDER WATER.

For many years this has been a great favourite; it is easily made in the following way:—Take 14 pounds of lavender-flowers; $10\frac{1}{2}$ gallons of rectified spirits of wine; and 1 gallon of water. Draw off 10 gallons with a gentle fire, or, which is much better, the sand-bath.

COMPOUND LAVENDER WATER.

Some persons much prefer this to the simple lavender water just spoken of. It is made thus:—Take of simple lavender water, 2 gallons; of Hungary water, 1 gallon; cinnamon and nutmegs, of each 3 ounces; red sanders, 1 ounce. Digest the whole three days in a gentle heat, and then filter it for use. Some add saffron, musk, and ambergris, of each half a scruple.

HUNGARY WATER.

Take of the flowery tops, with the leaves and flowers of rosemary, 14 pounds; rectified spirit, 11½ gallons; water, 1 gallon. Distil off 10 gallons with a moderate fire. If you perform this operation in *balneum mariæ*, your Hungary water will be much finer than if drawn by the common alembic.

This is called Hungary water, not because "Kossuth" came to this country, (who, by-the-by, would have done as well if he had stayed on the other side of the water,) but from its being first made for a princess of that kingdom.

SOME GENERAL DIRECTIONS FOR THE DISTILLATION OF SPIRITUOUS WATERS.

It has been ascertained that the plants and their parts ought to be moderately and newly dried, except such as are ordered to be fresh gathered. After the ingredients have been steeped in the spirit for the time prescribed, add as much as will be sufficient to prevent a burnt flavour, or rather more. The liquor which comes over first in the distillation is kept to itself, by some, under the title of "spirit," and the other runnings, which prove milky, fined down by art. But it is better to mix all the runnings together, without fining them, that the waters may possess the virtues of the plant entire; which is a circumstance to be more regarded than their fineness or sightliness. In the distillation of these waters, the genuine brandy obtained from wine is directed. Where this is not to be had, take instead of that proof-spirit half its quantity of a well-rectified spirit, prepared from any other fermented liquors.

In this steep the ingredients, and then add spring-water enough, both to make up the quantity ordered to be drawn off, and to prevent burning. By this method more elegant waters may be obtained than when any of the common proof-spirits, even that of wine itself, are made use of. All vinous spirits receive some flavour from the matter from which they are extracted; and of this flavour, which adheres chiefly to the phlegm or watery part, they cannot be divested without separating the phlegm, and reducing them to the rectified state of spirits of wine.

JESSAMINE WATER.

It is well known that there are several species of jessamine, but the sort intended in this instance is what gardeners call Spanish White, or Catalonian Jessamine; this is one of the most beautiful of all the species of jessamine. It is made as follows:—Take of Spanish jessamine-flowers, 12 ounces; essence of citron or bergamot, 8 drops; fine proof-spirit, 1 gallon; water, 2 quarts. Digest two days in a close vessel, after which draw off 1 gallon, and dulcify with fine loaf-sugar.

EAU DE BEAUTÉ.

The name of this water is taken from its use in washing the face and giving an agreeable smell. It is drawn from thyme and marjoram, which gives it a very elegant odour. Take of the flowery tops of thyme and marjoram, each 1 pound; proof-spirits, 5 quarts; water, 1 quart Draw off by means of a sand-bath till the feints begin to rise, and keep it close stopped for use.

FEINTS

SOME REMARKS ON THE USES OF FEINTS, AND THEIR GENERAL CHARACTER.

It will be observed that in the foregoing part of this work the receiver has been ordered to be removed as soon as the *feints* begin to rise, as the goods would otherwise contract a disagreeable taste and smell. It is not, however, to be understood that these feints are to be thrown away, nor the working of the still to be immediately stopped. Therefore, as soon as you can find the clear colour of the goods begin to change to a bluish or whitish colour, remove the receiver, and place another under the nose of the worm, and continue the distillation as long as the liquor running from the worm is spirituous, which may be known by pouring a little of it on the still-head, and applying a lighted candle to it; for if it is spirituous it will burn, but it will not otherwise.

When the feints will no longer burn on the still-head, put out the fire, and pour the spirits into a cask provided for that purpose; and when, from repeated distillations, you have procured a sufficient quantity of these feints, let the still be charged with them almost to the top; then throw into the still 4 pounds of salt, and draw off as you would any other charge as long as the spirit extracted is of sufficient strength; after which the receiver is to be removed, and the feints saved by themselves as before.

It may be remarked that the spirits thus extracted from the feints will serve in several compositions as well as fresh; but they are generally used in aniseed cordials, because the predominant taste of the aniseeds will entirely cover what they had before acquired from other ingredients. Such are the points to be taken notice of on this subject.

RULES FOR DETERMINING THE RELATIVE VALUE AND STRENGTH OF SPIRITS.

The following requisites are necessary to be obtained before this can be done in a satisfactory manner:—The specific gravities of a certain number of mixtures of alcohol and water must be taken so near each other as that the intermediate specific gravities may perceptibly differ from those deduced from the supposition of a mere mixture of the fluids; the expansions or variations of specific gravity in these mixtures must be determined at different temperatures; some easy method must be contrived for determining the presence and quantity of saccharine or oleaginous matter which the spirits may hold in solution, and the effect of such solution on the specific gravity; and, lastly, the specific gravity of the fluid must be ascertained by a proper floating instrument, with a graduated stem or set of weights, or, which may be more convenient, with both. They will be well suited for answering the purpose of the operator.

OBSERVATIONS ON DISTILLATIONS OF A SPECIAL CHARACTER, AND ON THE SELECTION OF APPARATUS MOST USEFUL.

THERE are numerous vegetables capable of furnishing elements for fermentation, and we may say that special distillations are equally as numerous. In another part of this work it must have been remarked that the nature of the wine operated upon, and the taste which it is necessary to give to the spirits, may command some particular mode of working to be followed, according as the tastes and flavours are to be removed or left in the produce. Still, if the system upon which the means of correcting or preserving are founded has been well conceived, it must infallibly have been remarked that these belong to the process of rectification; that is, to all mechanical operations the object of which is to give to alcohol a greater concentration.

It is evident that the object of improved apparatus being to effect the rectification of spirits with greater economy, in this respect the choice of apparatus is not attended with difficulty, and that such as present the most economical advantages must be preferred.

In fact, if the object of those improved systems is to produce at once spirits at the highest strength required, they are equally proper for the preparation of spirits at a lower standard.

It is thus that an improved apparatus may be used for the purpose of preparing three-six and proof goods with advantage; and if in the first case its economical superiority over simple apparatus is greater than in the second, its advantages in the latter case are of sufficient moment to render it preferable. The reader's attention will not be directed any further to the choice of apparatus; and there is justification in thinking that a sufficient and unexceptionable guide has been offered in regard to economical questions, if there did not exist a powerful consideration which may sometimes cause the distiller not to be influenced by the question of economy in the choice of apparatus.

For instance, such is that of the distillation of lees, as also that of grain, and of potatoes in the natural state. This, particularly, would cause an admission of a distinction in distillation, and consequently the latter will be divided into two kinds, the first of which will be called "distillation of fluid matter;" the second, distillation of half-fluid, half-solid matter.

Each kind of these distillations will be treated of separately, and to each of them will be assigned the apparatus and modifications that may be thought applicable.

In relation to the distillation of fluid matter, it is that which is effected on wines containing little or no original substances in suspension, although they may retain a more or less considerable quantity of the latter in dissolution; such are the wines of the grape, of molasses, of saccharified fecula, of beer, and other extracts of grain. These kinds of wine are those that offer the least difficulty to distillation. It may be effected in any kind of distilling apparatus; so that, in this case, that which is the most perfect may at once be chosen, without the least apprehension of inconvenience resulting from the state of the wine.

It is evident that nothing but some considerations depending on the fitness of the workmen could now prevent the distiller from making use of improved apparatus, which always requires more intelligence and more care than that which is less complicated.

There is no doubt, though, but that in all cases in which the spirits are to be drawn off at a high strength, or to be corrected by rectification, the distiller would find an advantage in being at the expense of employing intelligent workmen to conduct the process of a better machine.

The economy then in combustibles and in labour acquired by such machine would amply indemnify the manufacturer for the higher price occasioned by the employment of more careful and intelligent workmen, particularly if the distillery is of some extent.

In other cases, where proof goods are only made, where the wine operated upon is rich, and where taste and flavour are not to be corrected by rectification, in such cases it might be possible that the distiller would not find the same advantage in making use of improved apparatus, particularly in establishments so small as not to admit of the system of continuity; then a simple condensing apparatus would be sufficient: besides, these discussions belong more immediately to the distiller. The manufacturer having once determined upon the choice of the apparatus he means to use, he has only to combine his operations, so as to give to his produce all the qualities requisite for consumption, and to regulate his way of working according to the state of the wine he operates upon.

It is thus that each time the distiller wishes to give to his spirit as much of the flavour of the fruit and of the wine as possible, he should not distil it at a stronger degree than is required for consumption.

He should draw the spirit as high as possible in all other cases; and if such spirit, lowered with water down to proof, has not quality enough, this proof must again be submitted to the still, to be more concentrated. This operation will always be practicable with the continuous apparatus, because every thing in this system will tend to favour it.

It has been seen, in fact, in this system of distillation, that the only difficulty which presented itself sometimes was occasioned by the wine being so rich as to be unable to condense its own vapours; for, all things equal, the proportion of water should always be greater, according as the spirit is to be drawn off at a low strength.

But as it is important here to draw the spirit at the highest standard possible, whether one or two, or even three, operations are resorted to, it will be conceived that with respect to condensation it will be found here the most favourable in support of the operation. The contrary would take place if, in operating on too rich a wine by means of the continuous apparatus, the spirits were only to be drawn at 19° or 22°,* to preserve all the good quality.

---

* As the thermometers of Reaumur and Fahrenheit are occasionally referred to in the course of this work and others upon the subject of distillation, therefore, in order to establish a correspondence between them, and to convert the degrees of the former into those of the latter, multiply the degree of Reaumur by 9, divide the product by 4, and to

It is evident, then, that if water were added to wine, for the purpose of rendering its distillation possible, the greater the quantity of water that is added the more obnoxious this addition will be to the quality of the spirits. If a fermented liquor were distilled by the simple apparatus, and it were necessary to improve the alcohol by the operations of rectification, it is evident that this rectification must be effected by means of passing the spirits repeatedly through the still.

Let it be supposed that in a similar case the first rectification gives part of its products at thirty and some degrees; it would be advantageous to separate this portion of strong spirit from that which runs afterward at a lower degree.

The combustibles necessary for the boiling and vaporization of this alcohol, if it were brought back to the still with the feints, would be saved; in such case these feints are rectified separately. It is true, that in working in this way it is necessary not to give over after each operation, but to work continuously, because there is always, at each rectification, a quantity of spirituous liquor left which is too small to make one charge. In the beet-root sugar fabric of M. le Duc de Raguse, at Châtillon-sur-

---

the quotient add 32; the sum expresses the corresponding degree on the scale of Fahrenheit. Secondly, to convert the degrees of Fahrenheit into those of Reaumur, from the degrees of Fahrenheit subtract 32, multiply the remainder by 4, and divide the product by 9; the quotient will be the degree according to the scale of Reaumur; and so on for the rest. This little explanation will prove of very great service to the reader, not only as regards distilling, but in other things also.

Seine, where the molasses is submitted to distillation, they work nearly in the way which has just been spoken of, and that with the only view of improving the quality of the produce. The spirit which runs at a strength above 23° or 24° is separated from that which runs at an inferior degree; and these two productions, separately conducted, form two different qualities, proceeding from the same run, of which that which is obtained at the highest standard, and lowered down with water, is the best.

It is now easy to account for that variation in quality which belongs entirely to the influence of rectification.

In fact, beet-root molasses contains an essential oil which is disagreeable, or which, by its nature, favours the formation of empyreumatic oil in the act of distillation; an acid is thus formed in the fermentation, and these causes of defect in quality, it is known, are more or less removed according as the alcohol is more or less cleared from the water with which it is mixed in the wine.

The various substances which might be the object of special distillations are so numerous, and the proportions of alcohol they might render are submitted to such exceptions and such modifications, that it would be difficult to give an exact and complete index of them.

The residue or spent-wash of fluid matter is not applied perhaps to any use. The only substances which it might retain, besides some calcareous salts of little importance, are undecomposed sugar, a gummy substance, and more or less extractive matter. When speaking of fermentation, the process was indicated that is to be followed to deprive the spent-wash, as much as possible, of the sugar which it retains after the first operation, and to effect this to the advantage of the alcohol.

This mode, which is only practicable in distilleries in which the preparation of wine is continuous, would almost leave in the spent-wash the only substances which do not directly concur to the formation of alcohol; and in general this spent-wash is wasted on leaving the still.

However, it might be possible to turn it to advantage, in many instances, as manure; and if the acids which they retain did not suit the nature of the soil for which they were intended, they might be neutralized by means of lime. It is a fact that the organized substances which it retains would be most useful to vegetation.

It would be necessary to calculate, in such application, whether the effects of such a manure would sufficiently indemnify the farmer for his expenses in carriage and in labour which it would occasion: I am of the opinion that it would not.

Some remarks will now be made on the distillation of half-fluid, half-solid matter. Wines of a semi-fluid, semi-solid nature may be very numerous, though, in fact, they are less so than fluid wines. The most remarkable, and those which, by their importance, solicit a more particular attention, are lees or ground wines, worts of grain and of potatoes, which have not been mashed by extraction.

Every means of perfection applied to any of these wines is applicable to all of them, and in this respect we might generalize what will be said on this subject; but, on the other side, there is this difference, that the wines of grain and of potatoes may more easily and with greater advantage be transformed into fluid wines than lees.

This consideration will call forth the necessity of treating separately on the distillation of these wines, and on the apparatus suitable to them.

In regard to *lees*, it has already been seen that these wines proceed from the fermentation of the waste of the raisin, such as the stalks, skins, and kernels, with water, either resulting from wine with which they have already fermented, or proceeding from the separation of the must by means of the press.

The fermentable matter which this waste still contains in this state, particularly when it has already undergone fermentation, is evidently that which has been separated by the press, and which, being still enclosed in the cells of the fruit, has thus escaped alcoholic decomposition. This fact again proves what has been said before on the imperfection of the operation of pressing; and, indeed, if this operation could be executed with the same degree of practical perfection which is obtained in a great number of other manufacturing operations, the preparation of *piquette* and of lees-wines might, without prejudice, be neglected.

It is true, that in this case the distillation of grounds or lees could not be dispensed with; for, admitting even the perfection of the operations of pressing, it would be necessary to separate the alcohol which the grounds still contain in tolerably large quantities, when, after having fermented with the must, they are separated from it by the press.

But if the difficulty were thus not completely removed,

it would, at least, be attenuated in many instances. The difficulty which is attached to the distillation of lees-wines is the solid substances which they retain in a state of suspension.

These substances, which are denser than the wine, precipitate themselves to the bottom of the vessels in which they are deposited; and if these vessels are stills exposed to the direct action of the fire, they cause them to stick and adhere strongly to the bottom, where they burn and give birth to all the products of the combustion of organical bodies, among which the empyreumatic oil is in large quantities.

The influence of this oil on spirituous liquors is too well known. Several means have been imagined to prevent this accident. Experience has taught, for instance, that when lees-wines has gained the temperature of ebullition, and when vapour is formed in a continuous manner on the bottom of the still, its rising, occasioned by the ascensive agitation, is an obstacle to the precipitation of solid matters, and of course to their torrefaction.

This phenomenon is easily conceived, and it is presumed it is not in want of being further developed. In consequence of this observation, a vertical bar has been established in the centre of the still, and by these means a chain has been made to sweep the bottom of it.

However, it has been ascertained that this precaution is not always efficacious, and that during the distillation the workman might happen to be neglectful in alimenting the furnace, so as to maintain the still in a complete movement of ebullition; the solid matter, not being any longer suspended, precipitates itself to the bottom of the still, and provokes the accident which has just been mentioned. Many authors have proposed the *balneum mariæ* for the distillation of lees: this mode would be good with respect to its effects, if the question of economy would admit it.

It has already been shown why this system of distillation is really not admissible. By these means the empyreumatic taste would be avoided, but the taste of lees, which is not caused by torrefaction, as will soon be shown, would not be obviated at all.

It has also been proposed to transmit through metallic surfaces the heat of steam, but this mode has the same weak sides, with respect to economy, as the *balneum mariæ;* so it must entirely be abandoned. It would not be the case in the distillation of lees by mixed vapours, and this mode is, perhaps, the only one practicable to obtain from lees all the alcohol they can produce, and of preventing, at the same time, torrefaction.

It consists in placing the lees in a wooden vessel, but better in a metal one, in which they are to be heated by means of a steam-pipe, similar to that which establishes a communication between the two stills of Adam and Berard. To this effect, a steam-boiler, a still for the lees, a condenser, and a worm would be wanted in a continuous work; the lees would be brought to the boiling point in the condenser, and would offer the advantages attached to this disposition. The number of lees-stills might be increased to two, or even three, by making them of small dimensions and placing them one above the other; but this would be the utmost of complications which might, without inconvenience, be adopted in this kind of work.

The steam-boiler should be supplied constantly with water, and it must be perceived that in consequence of this exigence this system would require, in each operation, the combustibles necessary to boil the water requisite for the distillation of the lees; these lees are rendered poorer when heated, for vapour of water which fills this function can only produce this effect through its condensation in the mass, by uniting with it until the ebullition commences, when this vapour determines the analysis.

It is true that with three stills the expenses would not be so considerable; but evidently they would always be supplementary to those which are attached to the distillation of fluid wines by the same process.

This mode of distillation is thus recommended to those it concerns, if it were only to deprive the pressed lees, obtained by the means that will be indicated, from the alcohol which they retain after the operation of pressing. If more complicated apparatus were made use of for the purpose of distilling lees, such, for instance, which, like the continuous apparatus, force the wine through numerous circulations before it arrives to ebullition, it would be difficult, not to say impossible, to obtain good results; the solid substances would keep in the angles of the apparatus, obstruct the conduits, and present a vast number of similar difficulties, which experience gives us no hope of removing. The other mode which has been proposed for the distillation of lees is this:—It consists in assimilating these wines to those that are perfectly fluid, by first separating by precipitation all the liquid they contain, and by submitting the solid residues to the action of an energetic press.

This mode would be precious, and free from any objections, if the lees collected in the press did not retain, after this operation, a considerable quantity of alcohol, which could not be well extracted by distillation. To conceive the cause of this fact, it will be sufficient to consider the mode of acting of organical bodies charged with water or alcohol—fruits, for instance, that have been preserved in brandy; after a certain time these fruits imbibe the alcohol of the brandy and emit the water.

The cause of this phenomenon is not well known, but the fact exists, and has no doubt attracted the notice of the reader.

In fact, fruits preserved in alcohol have always a greater alcoholic taste than the liquor in which they have been preserved: this has been the cause of its being commonly said that "fruits drink spirits." The same phenomenon takes place in all wines which have fermented with solid substances: these contain always more alcohol, in proportion to their weight and volume, than the liquid in which they are formed.

When the solid substances of the lees are merely separated by the press, the production of spirits is considerably lessened by not submitting to distillation the substances which retain the alcohol in the greatest proportion: this fact has been verified by comparative experiments on grain and potatoes.

On the other side, the spirit thus produced gains much in quality, and the cause of this acquisition is easily explained by the results of numerous inquiries on that subject.

The distillation of the skins of the raisins, in which the

essential oil is seated which gives the lees taste, properly speaking, is, in fact, avoided by these means. Thus, it would be necessary, in the choice of the method to be followed in the distillation of lees, to discuss whether, on one side, the acquisition of quality obtained, with the loss of a certain portion of alcohol, is not more advantageous to the interests of the distiller than to obtain the whole of the alcohol, subject to the infectious taste of the lees of empyreuma, and, moreover, with the danger of all the difficulties attached to the distillation of half-fluid, half-solid substances.

It is thought that the first of these two propositions unites the most causes in its favour; and this opinion is the better founded, as it may be possible, by adopting the method which it embraces, to remove the only weak side which it presents.

Suppose a given quantity of lees transformed into fluid wines by separating the solid substances by means of the press: the fluid matter should first be distilled by the same apparatus and the same processes as wine, in the class of which these operations would thus place it; and, besides, the solid substances might be distilled by means of the steam of water, with the disposition which has been recommended as useful in treating lees.

It is thus that two qualities of spirits would be obtained—one of which would scarcely differ from that of fluid wines, and the other bearing all the taste of the lees, of which it might be freed by rectification. This mode conciliates sufficiently, it is thought, the chances of success to attract the attention of distillers of lees, and is in perfect harmony with the principles heretofore set forth.

With regard to the recommendation which has been made to saturate the acid of the lees with chalk, it is considered to be good; but the use of chalk is at an end there, it would seem, and does not in the least contribute to neutralize the essential oil. The solid substances of lees, when dried and burned, give a product which is called "lees ashes;" this operation is a true incineration, the products of which are gathered. Among these products, which are all of a calcareous nature, the tartrate acid of potash is found in large quantities, and it is to this body, useful to arts, that lees ashes owe their value.

It is very often that the residues of the distillation of lees are used as manure, and this agent of reproduction is tolerably appreciated in the vineyards.

In fact, it is a true consumption, in the place of production, which assigns to this mode of working all the advantages which it offers to science and to agricultural purposes.

The observations on this article will be brought to a close by giving an extract, made by M. Gay-Lussac, out of a memorial of M. Aubergier on the spirits of lees. This extract, which is taken from the "Annales de Chimie et de Physique," will give further information on what has been said previously concerning the special distillation of lees :—

"Until the present day, it has been thought that the flavour and the acid and penetrating taste of lees-brandies were owing to a certain oil, which, according to some, was formed during the process of distillation, and according to others, existed already formed in the kernels of the raisins

"According to the observations of M. Aubergier, it would appear that this oil is seated in the skin of the raisin itself, and, from the facts which he relates, his opinion is likely to be true. Kernels distilled with alcohol or water have given a liquor of an agreeable taste.

"The stalks have produced, by distillation, a liquor slightly alcoholized, having neither the taste nor the flavour of lees-brandy. But the envelope of the raisin, separated from the kernels and from the stalks, when submitted to distillation, after having been fermented, have given a spirit in all respects similar to that of lees. Thus it appears clearly demonstrated by these experiments that the seat of the oil, which communicates to the lees-brandy its bad qualities, is in the skin of the raisin. M. Aubergier has succeeded in obtaining this oil by rectifying lees-spirit at a moderate heat.

"The first portions of alcohol which came over had much less acridity than those that followed: on having been rectified a second time, they were almost entirely free from it; but repeated rectifications could not give it so agreeable a taste as that possessed by the spirit produced from wine. The latter portions of liquid in each operation, reunited and distilled, gave, at first, alcohol, which the addition of water did not render troubled, and which contained but little oil.

"The portions which were afterward obtained were transparent, but they became troubled when mixed with water; the third portion, which remained milky until the end of the operation, had on its face a light couch of oil, although it marked 23° Beaumé.

"This last produce having been mixed with the second,

and a suitable quantity of water having been added to bring them down to 15°, the liquor became immediately opacous, and a quarter of an hour after it was covered with a quantity of oil: 150 litres have produced more than 30 grammes of this oil. This oil has the following characteristics:—

"It is extremely limpid and colourless the moment it is separated from the alcohol, but the light gives it, a few moments after, a slight lemon colour.

"It is very fluid; its flavour is penetrating, and its taste very acrid and disagreeable. Submitted to distillation, the first portions that are volatilized keep their aroma; but the product soon acquires an empyreumatic taste, which, M. Aubergier suspects, is caused by a small portion of fixed oil proper to the kernel of the raisin; the liquor left in the retort takes at the same time the colour of lemon, which increases during the operations, and leaves at last a very light coal."

To the above, M. Gay-Lussac adds the following note:—

"It is not necessary, to explain this fact, to resort to the presence of a fixed oil in that which is drawn from lees-spirit; for the latter, although it has a very acrid taste and flavour, is nevertheless much less volatile than essential oils."

Then proceeds the subject thus:—

"It combines with water in the proportion of one thousandth part, and gives to it the particular flavour and acridity.

"When in ebullition it dissolves sulphur, which is precipitated by cooling, and with alkalies it forms soap.

"The oil is so penetrating and so acrid that one drop of

it is sufficient to infect 100 litres of the best brandy. M. Aubergier remarks, that the spirits that are drawn from the various fruits owe their particular taste and flavour to a volatile and oily principle, generally found in the surface of each fruit, and that, by taking this surface away, they would almost all be alike; that by thus depriving apples, pears, plums, apricots, peaches, and even *barley*, of their envelopes, spirits would be drawn from these vegetables almost entirely free from the flavour inherent to them."

To this M. Gay-Lussac adds a note as follows:—

"Many persons attribute the taste and flavour of lees-spirit to distillation itself, during which the lees stick to the sides of the still, which causes them to be carbonized.

"One thing which confirms the influence of this fact is, that when lees are distilled by the new process—that is, by the steam of water—spirits of a much better quality are obtained. However, it is not less certain that lees-spirits contain a peculiar essential oil, odorous, very acrid, altering their quality very much, and on which M. Aubergier has made interesting remarks. This oil, by its flavour, its acridity, and its property of not staining paper, and of not being converted into soap by alkalies, must be classed among the number of essential oils; but its property of being little soluble in alcohol, of burning without smoke, and of being much less volatile than the rest of the essential oils, which I have verified on the sample obtained by M. Aubergier, prove that it has some analogy with fat oils."

It may not be amiss to say a few words concerning the "semi-fluid, semi-solid wines of corn and potatoes." Every thing that has been said on the special distillation of lees, and on the apparatus suitable to it, is also applicable to the wines or worts of potatoes and of grain. To this case are again attached all the inconveniences attending its application to the distillation of lees.

In fact, the dangers of torrefaction depend, in this case, on the workmen, which is not a sufficient guarantee for their disappearance, and they may often be reproduced. The taste of empyreuma exists thus always more or less intensely in spirituous products, and these retain, besides, the bad taste and flavour of the fruits.

It would appear, in fact, as if these peculiar and distinctive variations did belong to an essential oil, which must principally reside in the husks of the grain and of the potatoes, and which is still incorporated in the spirituous produce more or less intensely, according as these substances have been introduced in smaller or larger quantities in the fermentation and distillation.

It may thus be conceived that the most efficacious correcting mode is that which has already been recommended, and which consists in fermenting nothing but very fluid extracts. It has not been the object, in discussing this matter, to introduce innovations prejudicial to the established mode of working distilleries which could not adopt them without injuring their interests: the point aimed at has been to signalize the causes of the different qualities of spirits, and the means of conquering them.

The condensing apparatus already mentioned is that which is most generally made use of in the distillation of corn and potatoes.

To a manufacturer who wished to establish a corn or

potato distillery in a country where a good quality of brandy is consumed, it would be useful to proceed in such a way as to give to the product the least taste of those vegetables possible; the object of working in this way would not be to identify the new liquor with that which is known and preferred, but it would, at least, be making a great step toward it.

The experience of the Parisian distillers is an instance of this case. In countries where large quantities of corn and potato spirits are distilled, some sort of essential oil is always incorporated with the liquor, which masks, if not the tastes, at least the peculiar flavours which the fruits and the various processes of distillation give to the produce.

The essential oil which is most generally used is that of juniper-berries; it is mixed in the still with the low wines in smaller or larger proportions, according as the spirituous product is to have a weaker or stronger taste of it.

This causes the corn spirit, of which so large a quantity is consumed in Belgium and in the North of France, to be called by the name of "geneva:" this name is given to the spirit even when it does not possess any aromatic flavour whatever. Instead of the juniper-berry, they often use other odorous substances, such as aniseed, wild oranges, &c., which are mixed with the low wines in the last rectification.

Similar means would thus contribute to give less utility to the various operations tending to improve the quality of the spirits. The distillation of grain and of potatoes is often combined with the feeding of cattle; and if, in some towns, there exist distilleries that do not dispose of their spent-wash in that way, it is because they can sell it to feeders, who in general hold it in great esteem.

Cattle are very fond of it; it nourishes well, and keeps them in a good state of health, if given in moderate quantities, and not too hot; but to use it as a powerful means of producing flesh, it should be mixed with linseed-cake, which makes the spent-wash a very effectual food. Oxen may by these means be well fed in three months, and they will look remarkably well indeed.

REMARKS ON AN INSTRUMENT INTENDED FOR TESTING WINES.

For a long time it was a desideratum at Paris to discover some easy mode of ascertaining the quantity of alcohol, or pure spirit, in wines destined for distillation.

Some time in the year 1810 a patent was granted by the French government for an *areometer*, that was to answer the intended purpose by being plunged into the wines that were to be tried, with an addition of carbonate of lime.

This, however, not bearing the test of experience, came to nothing, and it was found necessary to have recourse to distillation. M. Descroizilles therefore resolved to attempt the construction of a small alembic, heated by spirits of wine.

The point most embarrassing was the refrigeration, or cooling, necessary to condense the vapours. The common mode required a vessel larger than the whole of the new intended apparatus; in this, only a little water was wanting. However, this difficulty being got over, it was found practicable, with a small lamp, to obtain a sufficient quantity of brandy in the course of half an hour.

A glass vessel which served as a recipient, by a very simple operation, was filled with a mixture in arithmetical proportion with the different wines contained in a number of tuns of various capacities.

This instrument, which would admit of the distillation of even a glass of wine, and afford the product in half an hour, was found to be such that it might be repeated at pleasure many times in a day. It was observed that people who had orange-trees, and who could only collect a few of the flowers, had now an opportunity of amusing themselves in drawing distilled waters.

They had nothing more to do than to put the water into the little alembic, and then to lay the flowers upon the two gratings, across which the water was to pass in a vapour in order to be condensed in the receiver.

People might also make similar experiments in distilling rose water, mint, peppermint, &c. At the same time it was observed that a great number of exotic vegetables cultivated in green-houses contained volatile oils and aromatic qualities scarcely known till a short time since, because their leaves, flowers, seeds, roots, and barks were too small to be distilled in the ordinary manner; but, with this little alembic for the trial of wines, repeated distillations might be made at the different epochs of the growth of these articles, and their products duly estimated.

Besides, as physicians often recommend distilled waters, sometimes not to be had, some ounces were now obtained in an hour. Further, in any course of chemistry this little alembic could be mounted upon a table in an instant, around which the professors might be sitting, and easily afford its products in the course of a lecture, besides serving as a kind of demonstrator with the greatest despatch.

This apparatus in miniature, being constructed of the best tin, is of an agreeable form, and unites in itself all the facilities for the operation for which it is intended. It requires no wrapping in paper, no luting, &c.; all the joints, though, are very exactly closed, and few instruments are better adapted. Young persons who may have very little instruction may now indulge the wish to study the arts of distillation, perfuming, or the making of sweet waters, and of chemistry in general. Nearly the whole of the parts may be enveloped in linen cloth, in which they may be rolled up in a minute with as much ease as safety in securing them from coming in contact with each other.

They are frequently enclosed in an oblong sack, which in its turn is put into a cylindrical tin box, sixteen inches long and about three and a half in diameter.

Even the cover of this box is an essential part of the whole apparatus. The weight of the apparatus is not more than six pounds and a half, including a tin vessel full of alcohol.

SOME GENERAL DIRECTIONS FOR THE PREPARATION OF VARIOUS CORDIALS, COMPOUNDS, &c.

The perfection of this grand branch of distillation depends upon the observance of the following rules, easy to be observed and practised:—The artist must always be careful to use a well-cleaned spirit, or one freed from its own essential oil. For as a compound cordial is nothing more than a spirit impregnated with the essential oil of the ingredient, it becomes necessary that this spirit should have deposited its own. Let the time of previous digestion be proportioned to the tenacity of the ingredients, or the ponderosity of their oil.

Thus, cloves and cinnamon require a longer digestion before they are distilled than calamus aromaticus or orange-peel. Sometimes cohobation is necessary; for instance, in making the strong cinnamon cordial, because the essential oil of cinnamon is so extremely ponderous that it is difficult to bring over the helm with the spirit without cohobation. Let the strength of the fire be proportioned to the ponderosity of the oil intended to be raised with the spirit.

Thus, for instance, the strong cinnamon cordial requires a much greater degree of heat than those from lax vegetables, as mint, balm, &c.

Let a due proportion of the finest parts of the essential oil be united with the spirit—the grosser and less fragrant parts of the oil not giving the spirit so agreeable a flavour, and at the same time rendering it thick and unsightly. This may, in a great measure, be effected by leaving out the feints, and making up to proof with fine, soft water in their stead. These four rules, carefully observed, will render this extensive part of distillation very perfect indeed.

Nor will there be any occasion for the use of burntalum, white of eggs, isinglass, &c. to fine down cordials, for they will presently be fine, sweet, and pleasant tasted, without any further trouble. Cordials and compounds of various kinds are now made to suit the peculiar taste of almost every individual; the art has been brought to great perfection.

OF SOME OF THE PLANS RESORTED TO FOR ADULTERATING BRANDY.

It is truly lamentable to see how far men will allow themselves to be carried from the honourable and upright course which they should pursue, for the purpose of amassing wealth! It is well demonstrated in the case now under consideration, in which persons will put into brandy and other liquors such things as are poisonous, knowing the deadly influence it will exert on those who use it.

The first of these sophistications is performed by the addition of other fermentable matter to the must before the fermentation takes place, which increases the quantity in proportion to the increase of the spirit produced by the matter so added.

The quantity of ardent spirit being thus augmented in order to render it wholesome, it is therefore less corrected.

This kind of brandy is evidently inferior to the genuine, and in a certain degree recedes from those distilled spirits which are reckoned safe and wholesome. Another method is by adding spirits of malt, already distilled, to the wine or fermented must, these being the cheapest; but they must have been previously rectified for this purpose,. and indeed for making any palatable spirituous liquors whatever.

The depravity of this kind of brandy is still greater than the first, as it comes over in the still nearly as so much ardent spirit mixed with the brandy; and it will of course exert its noxious qualities on those who drink it.

Some persons adulterate brandy by the addition of simple rectified spirit or by counterfeit brandy; but the far most general method is by putting a counterfeit kind to the genuine.

This counterfeit brandy is made of malt spirits, first rectified, and then dulcified by redistillation of acids. The rectification of malt spirit, in order to make brandy, is always necessary, on account of its being impregnated with a proportion of empyreumatic oil in the first distillation, which oil is commonly called the "feints."

These give a very disgusting taste and smell to the spirits distilled. The substance much used for keeping down the feints is a medicinal preparation, called *lapis infernalis.* Its effects, with the redistillation, bring the ardent spirit to that state in which it abounds with noxious qualities, and, though it is freed of the feints, has a great effect upon its wholesomeness.

This *lapis infernalis* is made by adding lime to pearlash, potash, or any other vegetable alkaline salt, dissolved in water; then drawing the clear fluid, and evaporating it till a dry mass remains. The acid used in the preparation of counterfeit brandy is commonly called "spirit of nitre," or *aqua-fortis*, which, when combined with the rectified spirit, raises a flavour and taste much resembling those of brandy; but if a certain proportion of water be mixed with such brandy, a separation of the ardent spirit and acid immediately follows. The noxious effects of these on the health of those who drink this kind of brandy are frequently lamentable in the extreme, for it makes a complete wreck of their mental and physical powers; all of which blame is to be attached to those who adulterate the brandy for the purpose of becoming rich, though in doing so they make dreadful havoc of human beings, and those, too, who most of all others contribute to their success in business, for they consume it, being led astray by an evil passion.

But in regard to the effects of deleterious substances on the human system, I have spoken at length in another work, written by me, entitled "Detection of Fraud and Protection of Health," published in Philadelphia in 1852, to which the reader is respectfully referred.

PROCESS FOR MAKING LIME WATER.

This is conducted after the following plan :—Take 8 pounds of unslaked lime; put it into a pail or tub, and pour on it 3 quarts of water to dissolve it: in about an hour after add 3 gallons more of water, and let it stand for twenty-four hours. Then pour the fine off into a cask, and put a cock in it, and it is always ready for use. It is the impression of some persons that lime water is not healthy, but it is now pretty generally admitted that it is very good for many things in a medicinal point of view.

PROCESS OF MAKING SULPHURIC ETHER.

This very useful medicinal preparation, which is extensively used at the present day, is made as follows:— Take of oil of vitriol and rectified spirits of wine, each 32 ounces. Pour the spirit into a glass retort that will bear the sudden heat, and pour the acid at once upon it; mix them gradually and cautiously together by gently shaking the retort, and immediately distil by a sand heat prepared beforehand for that purpose, the recipient being placed in a vessel of snow or water.

The fire should be so regulated that the liquor may

boil till 16 ounces are distilled, when the retort is to be removed.

To the distilled liquor add 2 drachms of the stronger common caustic, and distil again, from a very high retort, with a very gentle fire, the recipient being placed, as before, in a refrigeratory.

Continue the distillation till 10 ounces are drawn off. To the acid residuum, after the distillation, if you pour 16 ounces of rectified spirit of wine, and repeat the distillation, more etherial liquor may be obtained, and this process may be repeated several times. The preparation of this singular fluid has long been confined to a few hands; for, though several processes have been published for obtaining it, the success of most of them is precarious, and some of them are accompanied with danger to the operator.

Where the dulcified spirit only is the object, the method as before directed for it succeeds to perfection; but when it is made with a view to the other, a variation is necessary, for only a small quantity of ether can be separated from the spirit so prepared. There, the distillation is performed with an equable and gentle heat; here, the fire should be hastily raised, so as to make the liquor boil, for on this circumstance the produce of ether principally depends. Ether is the lightest, most volatile, and inflammable of all known liquids.

It is lighter than the most highly rectified spirit of wine, in proportion of about 7 to 8. A drop let fall on the hand evaporates almost in an instant, scarcely rendering the part moist.

It does not mix but in small quantity with water, spirit of wine, alkaline lixivia, volatile alkaline spirits, or acids; but is a powerful dissolvent for oils, balsams, resins, and other analogous substances.

It has a fragrant odour, which, in consequence of the volatility of the fluid, is diffused through a large space. Its medicinal virtues are, too, well known; it acts on the nervous system with great power, rendering the person entirely insensible if taken by inhalation; and when taken into the stomach, in combination with other substances, has a soothing influence, easing pain and procuring rest.

It is an excellent remedy for headache, used in combination as follows:—

Take of Sulphuric ether.......... 1 ounce,
" Chloroform.............. ½ "
" Cologne................. ½ "
" Laudanum............... 1 tea-spoonful.

Some of this mixture should be put into a saucer or other suitable vessel; then linen cloths should be soaked in it and laid on the temples, or whatever part of the head may be affected.

It is also a pleasant disinfecting agent, used in the sick-chamber: cloths dipped in it and laid over any part where there is pain will have a grateful influence, temporary if not permanent, and thus give time for the exhibition of other remedies with a view to removing the cause or causes of the particular affection. It can be taken internally in doses of from 5 to 40 drops, according to the age of the patient or the exigency of the case; it should be taken in a mixture of wine and water, and swallowed imme diately after being dropped from the vial, as it exhales very soon after being exposed to the atmosphere.

INSTRUCTIONS FOR MAKING INFUSIONS, SPIRITUOUS TINCTURES, &c.

This constitutes an important part of the business for those who are engaged in distilling or otherwise dealing in spirits. Rectified spirits of wine is the direct menstruum of the resins and essential oils of vegetables, and entirely extracts these active principles from various vegetable matters, which yield them to water either not at all or only in part. It dissolves likewise the sweet saccharine matter of vegetables, and generally those parts of animal bodies in which their peculiar smells and tastes reside.

The virtues of many vegetables are extracted almost equally by water and rectified spirit; but in the watery and spirituous tinctures of them there is this difference, that the active parts in the watery extractions are blended with a large proportion of inert gummy matter, on which their solubility in this menstruum in a great measure depends, while rectified spirit extracts them almost pure from gum.

Hence, when the spirituous tinctures are mixed with watery liquors, a part of what the spirit had taken up from the subject generally separates and subsides, on account of its having been freed from that matter which, being blended with it in the original vegetable, made it soluble in water.

However, this is not universal, for the active parts of some vegetables, when extracted by rectified spirit, are not precipitated by water, being almost equally dissoluble in both menstrua. Rectified spirit may be tinged by vegetables of all colours, except blue.

The leaves of plants in general, which give out but little of their natural colour to watery liquors, communicate to the spirit the whole of their green tincture, which, for the most part, proves elegant, though not very durable.

TONIC AND ALTERATIVE CORDIAL.

This extremely useful and tonic compound, which is very cheap, and at the same time easy to be made, is prepared as follows:—Take of gentian-root, sliced, 2 ounces; Curacoa oranges, 1 ounce; Virginia snake-root, half an ounce; cochineal, 10 grains; French brandy, 2 pints. This must be let steep for three days, and then strain through a cloth.

Of this a table-spoonful may be taken, three times a day, in a wine-glassful of cold water. It strengthens the digestive powers of the stomach, increases and invigorates the appetite, and arouses the secretions generally, but more particularly that of the liver.

AROMATIC BITTERS.

Another preparation, somewhat similar to the one just spoken of, will here be given; it is superior to the former in several respects. There are substances in this compound which are intended to make it more agreeable to the taste, and at the same time more generous in its influence on the stomach. It is thus prepared:—Take of gentian-root, sliced, 2 ounces; columbo-root, bruised, half an ounce; bark of wild-cherry, bruised, 1 ounce; yellow rind of Seville orange-peel, dried, 1 ounce; cardamom-seeds, freed from the husk and bruised, half an ounce; French brandy, 3 pints. Digest without heat, and strain off the bitters.

A table-spoonful three times a day, just before eating, will have a wonderful effect on weak and debilitated constitutions.

PROCESS FOR MAKING A DIURETIC AND STOMACHIC COMPOUND.

By the term "diuretic," we mean a substance that will act on the system through the medium of the kidneys, and thus carry off much effete and offensive matter. This compound is made in the following manner:—Take a bottle that holds 2 or 3 quarts; put into it 2 pints of good brandy, and add the following seeds, well beaten together in a mortar: 2 grains of angelica, 1 ounce of coriander, 1 drachm of fennel, the same of aniseed, and 15 grains of juniper-berries; add to these the juice of 2 citrons, with the dried peels, and 1 pound of sugar.

The whole must be infused in the bottle for four or five days; and do not forget to shake it frequently during the time, for the purpose of melting the sugar and extracting the virtues of the seeds.

Afterward strain the liquor through a cloth, to purify; then put it up in bottles, and keep it well stopped, to prevent it from deteriorating. This preparation is useful in cases of vomiting, pain in the stomach caused by eating some article of food which did not agree with the person, colic, acidity of the stomach, difficulty of breathing, and various other little ailings of a similar character.

The dose is one or two tea-spoonfuls, taken in a wine-glassful of cold water.

PROCESS FOR MAKING TINCTURE OF MUSK.

Musk is a substance which, in itself, is peculiar; some persons are excessively fond of it as a perfume, while others cannot tolerate it under any circumstances—in fact, being so very offensive to them that they cannot remain in a room where it is. I suppose, though, that in some form or other, most persons are fond of musk for toilet purposes. The following directions will enable those who wish to use it to make a tincture which they can use to suit their own particular taste, as regards strength, &c. :—Take of musk and white sugar-candy, each 1 drachm; rub them well together in a marble mortar, adding by degrees, during the rubbing, 5 ounces of rectified spirit of wine. Put the whole into a matrass or flask, digest for three days in a gentle heat, and pour off the clear essence, which must be kept stopped close in a bottle.

This tincture may be added to the ordinary "cologne water," which will much improve its odour, or it may be sprinkled in the inside of trunks for the purpose of scenting the clothes; it can also be added to such toilet articles as may seem fit or desirable to the person using it.

APPENDIX.

PRACTICAL DIRECTIONS FOR DISTILLING.
FROM THE FRENCH OF TH. FLINZ,
BREWER AND DISTILLER.

PRELIMINARY OBSERVATIONS.

The object of distillation is to obtain from solid matters as corn, beet roots, potatoes, and vegetable products generally, a spirituous liquid; therefore it becomes important to establish precise rules by which we may arrive at the best results.

The crude materials are submitted to four essential operations:—

    I.—MACERATION;
    II.—FERMENTATION;
    III.—DISTILLATION;
    IV.—RECTIFICATION.

The art of distilling, known from time immemorial, is not in itself very difficult. However, experience proves that the operation is the more successful and lucrative as there has been more exactness applied during its various stages. Our duty is therefore to point out with exactitude the best way to obtain the most advantageous percentage.

With this object, we shall examine the different operations of the distiller, and shall dwell on those points which are often neglected or overlooked in distilleries.

We use the Centigrade thermometer for our indications of temperature.

PART FIRST.

I.

MACERATION.

Maceration consists in submitting a solid body to the action of a liquid, in order to disengage one or several of the constituent principles of the primitive substance.

Generally, the solid substance is a mixture of rye and malt (we shall give the preparation of malt further on); and the liquid is water at a certain degree of temperature. The object is to produce the saccharification, that is to say, the formation of the saccharine principle.

We have said "a mixture of rye and malt," although the term "fleshy fruits" would be more correct. Nevertheless we keep the word rye, because on account of its relatively low price, it is the most generally employed.

Exactness in the mode of operation requires all our attention.

It is needless to say that, in order to work advantageously, it is absolutely necessary that the vessel, beck, vat or tun should be clean and in good order.

Modus operandi. Take one part of malt and four of rye, finely ground, generally we use 15 to 17 Kilogrammes— 33 to 37.5 lbs. avoirdupois—of the ground mixture per hectolitre—3.5 cubic feet—of the capacity of the vat, in winter; and from 12 to 13 kilog.—26.5 to 28.6 lbs.—in summer. These proportions, of course, are approximative and should be modified according to the temperature, the season, and the experience of the operator. The grain and malt should be well mixed. Put this mixture into the vat where there is already about 1.5 litre or kilogramme of water—1.5 quart for each kilogramme—about 2 lbs.—of the ground mixture. The temperature of the water is from 60° to 70° C.—140° to 158° Ft. This quantity of water is also an approximation, because the power of absorption varies greatly with the various kinds of flour. The operator can add as much water as he thinks necessary, provided that the paste be firm and consistent, and at the same time, thoroughly moistened. When the mixture of water and flour is completed, its temperature is about 40° C.— 104° Ft.

Allow the mixture to stand from 15 to 20 minutes, during which time, lactic acid will be formed; then add boiling water in the proportion of about half a litre (1 pint) for each kilogramme (about 2 ℔s.) of the first mixture. When the paste has become thoroughly mixed, again add boiling water in the proportion of about 1 to 1.5 litre (1 to 1.5 quart) for each kilog. (2 ℔s.) of flour.

At each addition of water, the operator must be careful to note the temperature of the mixture, which should remain between $60°$ and $67°$ C. ($140°$ and $152°.6$ Ft.) and never above $70°$ C. ($158°$ Ft.), for, should it exceed the latter temperature, the operation would not be successful on account of the too rapid formation of sugar.

After the third addition of water, the vessel is covered, and the whole allowed to rest for about fifteen minutes, in order to obtain a uniform temperature in the mass; then, the mixture is thoroughly worked for five minutes, and again left covered for about half an hour, so as to allow the saccharification to become perfected.

Afterwards, the whole mass is cooled by the addition of clear spent wash, (distilled residue.) A draught of air is made to pass through the room, and, if there is a ventilator, cold air is blown upon the surface of the mixture during the operation. The temperature of the paste is to be decreased in a ratio with the temperature of the liquids which are intended to fill the vats. If the liquid, whether spent wash or water, marks $10°$ C. ($50°$ Ft.), the temperature of the paste should be lowered down to $50°$ C. ($122°$ Ft.); if the former marks $15°$ C. ($59°$ Ft.), the latter will have to be put down to $45°$ C. ($113°$ Ft.); and so on, for each higher degree of the liquid to be added, the temperature of the paste is to be previously lowered one degree.

Care must be taken if it requires a work of twenty-four hours, to keep the whole at a temperature between $30°$ and $31°$ C. ($86°$ to $88°$ Ft.), for, should it be below $30°$ C ($86°$ Ft.) the fermentation will not be perfect; and, on the other hand, a temperature above $31°$ C. ($88°$ Ft.) will produce a large loss of carbonic acid gas, and there is danger that acetic fermentation will occur.

The vat having been filled, and the mixture being at a temperature of $31°$ C. ($88°$ Ft.), the mass is stirred until the paste is uniformly and thoroughly distributed. Then we add one-fourth or half a kilogramme (0.5 to 1 ℔.) of brewer's yeast, or dry yeast of Holland, or from two to three litres (2 to 3 quarts) of artificial yeast per hectolitre (3.5 cubic feet) of the capacity of the vat.

II.

FERMENTATION.

As this operation is mysterious in its action, and begins and ends by itself, when the preceding maceration has been properly conducted, we shall confine ourselves to the enumeration of the various kinds of fermentation. At the same time, we shall indicate the characteristics by which we may ascertain when the operation proceeds well or the reverse.

There are five kinds of fermentation:

Foaming fermentation; Vinous fermentation; Alcoholic fermentation; Acetic fermentation; Putrid fermentation.

We shall briefly examine each of these:—

I. Foaming fermentation, also called saccharine fermentation, transforms into sugar the substances which during the maceration were not thoroughly converted into sugar. The operation will require a greater or less length of time according to the more or less complete saccharification during the maceration.

II. Vinous fermentation is the stage of transition between foaming fermentation and

III. Alcoholic fermentation, which transforms into alcohol the product of the preceding fermentations.

Alcoholic fermentation is thorough when the densimeter of Gay-Lussac marks $0°$.

In Belgium, the liquors seldom mark as low as $0°$ after twenty-four hours; there always remains a certain proportion of undecomposed sugar, and the distillation may be proceeded with when the densimeter marks $5°$.

If the fermentation were allowed to continue long enough for the densimeter to indicate $0°$, there would be imminent danger of

IV. Acetic fermentation, the result of which would be the production of vinegar, and which itself may give rise to

V. Putrid fermentation, the name of which is sufficient to indicate the result.

This putrid as well as the acetic fermentation, instead of succeeding the alcoholic fermentation, may, by want of cleanliness of the vat and utensils, or by an impure spent wash, occur immediately after or simultaneously with the vinous fermentation. This always occasions great loss, not only in the quantity, but also in the quality of the products.

When the fermentation is at its height, we must prevent the matter running over. For this purpose, take one part of lard and one of green soap, and with this mixture smear the inside and top edges of the vat; at the same time touch the top part of the scum or foam with a wooden stick smeared with the above mixture. Nevertheless, we must use this remedy only when there is danger of running over, because we may injure the fermentation by stopping it.

CONDUCT OF THE FERMENTING OPERATION.—Four or six hours after the yeast has been put into the vat, a crust or cap appears at the surface and cracks at several places. At this moment the vat must be uncovered. The foaming fermentation has just begun. More and more openings appear in the cap, and a sort of motion takes place in the vat, while more or less muddy bubbles appear.

We may judge how the saccharification proceeds by the color of these bubbles: are they plainly grayish white, then the operation is incomplete; on the other hand, a clear and white color is a sign that the saccharification is too much advanced.

This fermentation lasts from four to six hours, the length of time depending, however, on the degree of saccharification already attained by the previous operation.

The vinous fermentation immediately succeeds the foaming fermentation. We must note, however, that, when the maceration has been well conducted and the saccharification nearly completed from a pure mash made in good proportions, the foaming fermentation will not take place, and the vinous one will be the first seen, thus producing a saving of three to four hours time.

The vinous fermentation is manifested by a production of bubbles smaller than the preceding ones, and yellowish in color. This fermentation is more tumultuous than the former, and crackling noises are heard in the vat. At the same time Carbonic acid escapes, and the production of this gas is the more rapid as the temperature of the liquid is greater. This vinous fermentation lasts about six hours.

It is at the beginning of this fermentation that we must prevent the contents from running over.

The operation proceeds well when a cap or crust is formed on the sides of the vat.

During the vinous fermentation there are generally three or four tumultuous motions by which the contents may run over. Then another crust is formed at the surface, of a thickness of from two to five centimeters (1 to 2 inches), and even exceeding that. This crust indicates the end of the vinous and the beginning of the alcoholic fermentation.

There is sometimes a production of alcohol during the vinous fermentation, in which case, this becomes blended with the alcoholic fermentation. The latter lasts about six hours.

Since so much time is required for the various fermentations, it becomes important that time should not be lost during the maceration.

The end of the alcoholic fermentation is indicated by the cap or crust diminishing in thickness, then falling to the bottom, and the liquid becomes clear and calm at the surface. By inhaling the vapors, a vinous and sharp sensation is felt.

It is necessary that the vat should be covered during the whole of the alcoholic fermentation.

The fermenting substances becoming naturally heated during the operation, we will remind our readers, that their temperature, at the beginning, is best ranging between $30°$ and $31°$ C. ($86°$ and $88°$ Ft.).

In a well conducted operation, we will obtain from 8 to 8.5 litres (2.11 to 2.24 gallons) of 50 per cent. of alcohol per hectolitre (3.53 cubic feet) of liquid in the vat, or 58 to 60 litres (15.32 to 15.85 gallons) of the same alcohol per 100 kilogrammes (220.55 lbs.) of mixed flour.*

III.

DISTILLATION.

This operation consists in separating the alcohol from the substances which are not volatile. To effect this, we use

---

* The gallon is the U. S. gallon of 231 c. inches; and the 50 per cent. alcohol contains 50 volumes of pure alcohol for 50 volumes of water.

an alembic heated, whether by the direct action of the fire, or by steam, or what is still better, a distilling column which always requires steam. The latter apparatus gives a great economy of time, and with it we need not fear the burning of the substances at the bottom of the alembic, which often occurs when the fire is directly applied.

Another advantage of the distilling column is that we avoid the agitation of the matters, otherwise necessary, until the whole is at the point of ebullition, in order to prevent the burning. This last occurrence, indeed, not only diminishes the quantity of the products, but also their quality, by imparting to them a disagreeable taste, which it is difficult to dispel.

We therefore recommend a distilling column made of from 13 to 17 compartments, where the steam enters at the bottom, while the wash (beer) is introduced at the top.

The operator will begin by heating the empty column with steam, until the condensed water runs out in a stream of the size of a quill. Then, the wash or beer is introduced at the top by means of a forcing pump.

In those works where the operator has not at his disposal the apparatus for direct distillation and rectification, the best products are obtained when the cocks for the introduction of the wash and of the steam are so regulated that a constant stream of low wines marking from 30° to 35° of the alcohometer,* is obtained.

On the other hand, with apparatus intended for distilling and rectifying at the same time, it may be advantageous to produce high wines marking from 55° to 70° of the alcohometer.†

IV.

RECTIFICATION.

This operation is intended to remove all the essential oils and foreign substances remaining in the phlegms, and thus to obtain the alcohol in the greatest state of purity.

We employ an alembic heated by the direct action of

---

* The alcohometer mostly used in Belgium and France is that of Gay-Lussac, giving the degree or percentage of pure alcohol in volumes. There is very little difference between the alcohometers of Tralles and Gay-Lussac.

† The low or high wines of the first distilliation, and which need to be rectified, are sometimes called *Phlegms*.

the fire, or preferably by steam circulating through a coil. In the latter case the temperature is easily regulated.

It is advantageous to mix the phlegms to be rectified with one or two litres (1 to 2 quarts) of oak charcoal, recently burned, and, if the charcoal has been left exposed a long time, to submit it to a red heat in a closed vessel, and, when cold, to pulverize it.

Boneblack or animal charcoal is better than wooden charcoal.

We may also add to the phlegms a half litre (1 pint) of freshly bruised juniper-berries.

These proportions of charcoal and berries correspond to three or five hectolitres (10.59 to 17.65 cubic feet) of *Geneva* liquor, to be obtained, according to the taste and aroma of the phlegms.

The same as for the distillation proper, the operation is begun slowly, and the heat gradually raised until the liquors run off in a regular and continuous stream.

The first runnings have generally an ethereal odor, are not clear, and their taste is disagreeable; therefore, the operator must collect, according to the size of the apparatus, the first seven to twenty litres (7 to 20 quarts) running out, which must be added to the phlegms or wines of the next operation.

He should do the same at the end of the operation, when the product marks only 45°, because a bad taste will again appear. All the liquors below 45° are also kept for a subsequent operation, and the rectification is ended when the products mark only 11° or 10° of the alcohometer.

The apparatus where the distillation and rectification proceed at the same time presents the advantages over those where these two operations are separated, that in the latter case there is always a loss of two to five per cent., and that by rectifying directly, the running liquors may be obtained at the same degree, let us say 50°; whereas by a separate rectification, the products mark too high a degree at the beginning, and run low at the end of the operation. It becomes, therefore, necessary to add a certain quantity of water in order to reduce it to the standard of 50°, which practice is very often injurious to the taste of the liquors

PART SECOND.

SPECIAL OBSERVATIONS.

I.

BUILDINGS.

When it is desired to establish a distillery we must be careful to choose a healthy place with plenty of free space around it; and if it becomes necessary to use old buildings, it is absolutely necessary to drain and purify the place, so that, in the future, no cause of local unhealthiness may hinder or disturb the operations.

The buildings themselves should be so constructed or transformed, that it will always be possible to allow draughts of air to circulate through the works.

The disposition of the various parts of the building must be both compact and suitable; that is the boilers, the machinery, and the distilling apparatus be near each other in the central portion; the fermenting vats on one hand, and the mills and store-rooms on the other, to form the sides.

In well constructed works, the rectified *Geneva* runs directly from the cooling-worm into casks, barrels or other special reservoirs in the store-room, so that, without pumping it again, it may be put directly into the barrels intended for delivery.

The distilling column and the cooling-worm should be five metres (about 16 feet) above the ground. With this arrangement, a syphon is employed to draw off the spent wash or slop into a settling vat, from whence the clear liquors may be decanted, while the solid residue is allowed to fall into other reservoirs to be used for feeding cattle.

The clear and settled spent wash is easily decanted into other coolers by means of spouts, and, at the proper time, is used in the macerating and fermenting vat.

Such an arrangement of vats renders only two pumps necessary:—One for water, the other for the beer—while in other distilleries where this disposition is not followed, eight or even ten pumps are employed.

II.

UTENSILS.

As a rule, all the utensils which belong to the distillery should be made of copper, to insure cleanliness in the operations. Iron is objectionable on account of the rust.

It is needless to add that the greatest cleanliness in everything is indispensable in distilleries. Therefore the copper utensils should be the subject of constant attention. The cooling vats are preferably made of copper. But when wooden vats are employed, and the works are in operation, it is sufficient now and then, to whitewash the inside of the vats with a thin milk of lime, which is allowed to remain for half an hour, and is afterwards carefully rinsed out with pure water. When a stoppage occurs, the inside of the coolers and other vats receives a thick coat of lime, and the vessels are also filled with water which is renewed every ten days.

A strict observance of these rules of cleanliness will secure the operator from the occurrence of acetic and putrid fermentations.

We also recommend the washing, now and then, of all the wooden utensils with a solution of bicarbonate of soda, which neutralizes all the acids impregnating the wooden substances. This same solution is also advantageous for cleaning those recesses and parts of the works which are difficult of access, and are, therefore, neglected in the ordinary washings.

III.

MACERATION.

Those substances, in which the sugar is, if we are allowed the expression, kept suspended in a free state, such as honey, molasses, and the juice of beet-roots, carrots, plums, apricots, etc., require no maceration. It is sufficient to dissolve the sugar in water hot enough to obtain a temperature of $30°$ to $31°$ C. ($86°$ to $88°$ Ft.), which is necessary for starting the fermentation for an operation which requires 24 hours.

Water is added in sufficient quantity to have the whole mass marking $5°$ of the hydrometer for syrups, which corresponds to 50° of the densimeter of Gay-Lussac. Indeed the most dense liquid which may be distilled. that is to say water saturated with sugar, marks 40° of the hydrometer for syrups, or 400° of the Gay-Lussac's densimeter.

The juice of fruits, beet-roots, carrots, plums, apricots, etc., may be obtained in two different ways:—

(*a*.) The raw fruits are rasped, and then pressed.
(*b*.) After boiling in water, the fruits are pressed.

The maceration of potatoes requires a previous steaming of the tubercle, which is then mashed in an appropriate machine. The malt is put first into the vat, in the proportion of one part of malt to four of potatoes; then the latter, just mashed, are added slowly and gradually so as to give time to the man who manipulates the mass, to make a thorough and intimate mixture. The remainder of the operation is exactly as we have explained for the maceration of corn and rye.

Beets, carrots, etc., may, as regards the maceration, be treated like potatoes, but the proportion of malt is much smaller.

IV.

FERMENTATION, DISTILLATION AND RECTIFICATION.

For cooling and diluting the substances in course of maceration, we employ clear spent wash, (clear part of slops,) the residuum of beer, the water of breweries, or, if none of these liquids are at our disposal, pure water.

Spent wash is never used unless in admixture with one half, or at least one-third, of pure water.

There are two principal reasons why we prefer the spent wash to the other liquors; first, because having absorbed the oxygen of the air it helps the fermentation; second, because it marks generally several degrees of the densimeter, which shows that it still contains a certain quantity of sugar, which is thus put to account. Experience also proves that the starch suspended in the spent wash helps the fermentation.

The fermentation of syrup, honey, and of saccharine substances, in general, does not differ from that of corn, the theory of which has already been explained.

When for the distillation proper, we introduce into the

alembic or the distilling column the fermented substances, these should be previously well stirred, in order to obtain a uniform mixture; that is to say, we should endeavor to have the solid portions, which have a tendency to fall to the bottom of the vessels, kept as much as possible, in suspension in the mass of liquid.

If the rectified products are intended to be sold as pure alcohol, the first and last runnings, as we have already said, should be kept apart, until a sufficient quantity has been gathered for a special rectification. But, there again, the last runnings will have a bad taste; therefore, we must be careful to collect all the liquids with an objectionable flavor and which can be sold for the manufacture of varnishes.

V.

YEAST.

Brewer's yeast, when new, is preferable to that of Holland, because the former contains certain principles or acids which aid the fermentation.

These active principles are generally to be found in much greater quantity in the vegetables of northern countries than in those of southern regions; and it is one of the reasons why the fermentation is more easy and rapid in northern than in southern countries, if comparative experiments are executed with local products.

If it is difficult to obtain sufficient brewer's yeast, we may employ Holland yeast. But, in order to obtain results as advantageous as with the former, we recommend to add, for each kilogramme (about 2 lbs.) of Holland yeast, one litre (about 1 quart) of a decoction of hops, made by boiling one kilogramme (2 lbs.) of hops in ten litres (10 quarts) of water for five or six hours. The evaporated water must be replaced by the addition of water, in order always to have the same quantity of water boiling.

This mixture of yeast and decoction of hops may be made two or three days in advance.

This method of treating Holland yeast presents many and important advantages: the yeast may be kept for several days; the strength of the fermenting principles is increased, and the essence of hops imparts to the product a very pleasant aromatic taste.

If these two kinds of yeast cannot be had, the distiller may substitute an artificial one prepared as follows:—

One part of malt and two parts of wheat flour are allowed to macerate, and to this mixture we find it advantageous to add a small proportion of hops.

We follow, on a small scale, the method which we have explained for the maceration; but we will operate at a temperature somewhat lower, from 5° to 10° C. (9° to 18° Ft.). The mixture itself ought to be more consistent, and the thicker, the better. When the paste is thoroughly mixed, it is allowed to stand about 15 minutes, then it is worked occasionally, without adding any liquid, until its temperature has been lowered to 35° C. (95° Ft.); after which the paste is worked no longer, and is kept in an uncovered vessel in a moderately hot room.

After twelve or twenty-four hours, fermentation sets in spontaneously, and lasts from two to three days, according to the climate and the quantity of matter operated upon. The whole mass has been transformed into yeast when the fermentation appears to have done its work.

The proportion of artificial yeast is about four or five times that of brewer's yeast. However, there is nothing absolute in this proportion which will be modified by the operator according to his own experience.

In order to save time, and when it is necessary frequently to prepare artificial yeast, a portion of the yeast already made may be used to hasten the fermentation of the new mixture.

VI.

MALT.

The preparation of malt consists in the artificial germination of barley. A vat, half filled with water, receives the barley which is well stirred so as to allow the bad grains to raise and float on the top, where they may be removed with a skimmer.

In summer, the water must be changed every twelve hours; in winter once in twenty-four hours is sufficient.

The barley is sufficiently softened and penetrated by water after twenty-four to thirty hours in summer, and forty-eight to sixty hours in winter.

In order to ascertain if the barley is sufficiently steeped, we take a grain of it; if it bends easily under the nail with out breaking we may consider it as being in a proper state. The water is removed from the back, and the steeped barley is allowed to remain there for four or six hours in summer, and ten or twelve hours in winter. The barley is then removed from the back and spread on the malting floor in layers of 10 centimeters deep (4 inches) in winter, and only 5 centimeters (2 inches) in summer. In winter it is shovelled every twelve hours, and in summer, every six hours. The barley is thus worked until it begins to present a point (thus showing the phenomenon of germination, when the embryo sets forth two germs or roots), at which time it is spread in layers of 7 to 8 centimeters thick (3 inches) in warm weather, and 10 to 12 centimeters (4 to 5 inches) in cold.

When by putting the hand into the layer, we feel a temperature of from $20°$ to $25°$ C. ($68°$ to $77°$ Ft.), and more firmness of the grains, and when, simultaneously, a kind of dew appears at the surface of the layers, it is time to shovel the grain, taking care that the grains on the inside should take a new position on the top or bottom of the layer, and conversely. At each shovelful, the operator must spread the barley as much as possible, thus putting each grain in contact with the air which favors the continuation of this artificial vegetation. Seven or eight hours after this first operation, the new layer presents the same phenomena, and requires another turning over. This is done three times, after the same signs have presented themselves.

The layer of the third operation is allowed to rest for six to twelve hours, according to the temperature.

The barley will have acquired all the qualities of malt, when, by opening the back part of a grain, we find inside the vegetable germ having three-fourths the length of the grain itself. A superior product presents also five or six filaments which are each twice as long as the grain.

The grain is then spread out in a well ventilated place, and part of it may go immediately to the malt-kiln. What is waiting to be dried is spread and turned at least twice a day.

The malt kiln should be heated slowly at the beginning, in order not to produce a horny malt, the sugar of which is dissolved with difficulty. The portion of malt in course of drying should be turned upside down and conversely, every hour, or at least every other hour. When the grain is easily broken, and leaves by friction streaks like those of chalk, then the grain is sufficiently dried.

We will remark that, if the malt is not intended for immediate use, the greater the lapse of time before its employment, the longer should it be dried; and in such case, the temperature of the kiln may be raised to from 40° to 45° C. (104° to 113° Ft.).

After being kiln dried, the malt is bruised by some mechanical contrivance in order to break the germs and separate them from the grains.

If we desire to use the malt immediately, we may separate the germs while the malt is still warm; but if it is to be kept for a certain length of time, the germs may be left, to be removed only before use.

A malt intended for distilleries will be found sufficiently good when, before being kiln dried, the vegetable germ has only half the length of the grain itself.

Instead of barley, rye, wheat, oats, etc., may be employed for the manufacture of a malt intended for distilling purposes.

VII.

PRESERVATION OF SPIRITUOUS LIQUORS.

Every one knows that spirituous liquors of a certain age are more highly esteemed than those recently made. Therefore, when an old product is scarce, it becomes advantageous to impart to a new one the qualities of the former, that is to say, an artificial ageing which cannot be distinguished from the true one.

For this purpose, instead of keeping the liquors in cisterns, which are generally made for saving room, they must be put into barrels or casks. The wooden staves remove the essential oils which impair the flavor, and the operation is aided in the following manner:—

The filled barrels are put into a room, the temperature of which is raised to from 20° to 30° C. (68° to 86° Ft.). When the liquid has reached this temperature, the room and its contents are allowed to cool off. This heating and cooling is repeated, even three to five times. In this manner, we obtain in the course of a fortnight, a product similar in quality to that which has been kept one year in store-rooms.

The bung hole of the barrels remains open during the whole time of the operation, and the loss occasioned by this mode of working is equal to that suffered by one year of ordinary storage, that is to say nearly two per cent. of the whole.

VIII.

RAW MATERIALS.

It is generally acknowledged that the grains harvested in Champagne are better than those of Africa, and in general, that the productions of the north are, for our purpose, preferable to those of the centre and south. The products harvested on sandy and light soils are better than the corresponding ones grown on rich ground; so much so, that with equal weights, experience proves that the former give a product from 5 to 10 per cent. greater than the latter, and this is equally true for distilleries or breweries.

The cause is due to an active principle which favors fermentation in distilleries, gives a better taste and flavor to beer, and renders it more easy to keep.

INDEX.

	PAGE
Acetic Fermentation	202
Adulterating Brandy	188
Advantages of Continuous Distillation.	36
Alcoholic Fermentation	202
Alembic	46
Alterative Cordial	195
American Apparatus	44
Aniseed Cordial	157
Apparatus for Distillation	17, 40
Apparatus in American and English Distilleries	44
Apparatus, Selection of	165
Appendix	199
Apple Brandy	135
Areometer	184
Aromatic Bitters	196
Arrack	124
Aubergier on Spirit of Lees	179
Balneum Mariæ	46, 174
Barley, Advantage of	76
Beet Rasp	128
Beet-Root Molasses	170
Beet-Root, Spirits of	127
Bitters, Aromatic	196
Blackamoor's Head	47
Brandies, To prevent Deterioration of	95
Brandy, Adulteration of	188
Brandy, Cherry	146
Brandy, Distilling, etc.	93
Brandy, Raspberry	147
Brandy Shrub, Process for Making	145
Brewer's Yeast	210
Brewing Hollands Gin	101
Buildings for Distilling	207
Capital	47
Caraway Cordial	153
Cellars	9
Chaff	77
Charge of a still	12
Cherries, Spirits of	133

	PAGE
Cherry Brandy, Process for Making	146
Cider Spirits	135
Cinnamon Cordial	151
Cinnamon Water	158
Citron Cordial	150
Coals	75
Coloring Spirits	143
Common Process of Malt Distilling	91
Compound Lavender Water	160
Compounds, Distillation of	187
Condensers	17
Condenser, Wine-warming	30
Conduct of Fermentation	203
Continuous Distillation	26
Cooler	31, 47
Cordial, Aniseed	152
Cordial, Caraway	153
Cordial, Cinnamon	151
Cordial, Citron	150
Cordial, Lovage	150
Cordial, Peppermint	152
Cordial, Tonic and Alterative	195
Corne d'Aboundance	27
Corn for Distilling	64
Cucurbit	47
Damask-Rose Water	158
Description of a Distillery	9
Deterioration of Brandies, Prevention of	95
Difficulties in Distilling	30
Directions for Cordials, etc	187
Directions for Distilling	199
Directions to a Distiller	11
Distillation	17, 204, 209
Distillation, Continuous	26
Distillation of Common Gin	106
Distillation of Molasses	140
Distillation of Rum	137
Distillation of Simple Waters	155
Distillations, Special	165
Distiller, Directions to	11
Distillery, Description of	9

INDEX.

	PAGE
Distillery, Fire in	13
Distilling Brandy	93
Distilling Column	27
Distilling, Directions for	199
Distilling, Malt	91
Diuretic Compound	196
Double Distilled Rum	139
Drying	74
Dulcifying	15
Dutch Geneva	98
Eau de Beauté	162
Eau de Luce	147
Egg-plate	18
Elder Juice	145
Empyreumatic Oil	173
English Apparatus	44
English Method	81
English Vinegar	154
Ether, Sulphuric	191
Explanation of Egg-plate	18
Fecula, Separation of	116
Feints	12, 24
Feints, their Uses, etc.	163
Fermentation	70, 84, 202, 209
Ferments	84
Fining	14
Fire in a Distillery	13
Flavoring Spirits	143
Fluid Matter, Distillation of	166
Foaming Fermentation	202
French Method	79
French Noyau	151
French Process of Distilling and Preparing Brandy	93
French Vinegar	153
Fuel for Drying	75
Gay-Lussac on Spirit of Lees	181
General Directions for Cordials, etc.	187
Gin, Common	106
Gin, Hollands	101
Grain used in Distilling	63
Grapes	93
Gravity of Worts	11
Head	47
Hippocrates' Bag	16
Hollands	98
Hollands Gin	101
Holland Yeast	210
Hungary Water	160
Imperial Ratafia	149
Improved Apparatus	37
Inequality of Heat Prevented	58
Infusions	194
Instructions for Making Infusions, etc.	194

	PAGE
Instrument for Testing Wines	184
Instrument to Prevent Inequality of Heat in Distillation	58
Irish Usquebaugh	148
Jamaica Rum	139
Jessamine Water	162
Juice, how Obtained	209
Kirsch-Wasser	133
Lapis Infernalis	190
Lavender Water	159
Lavender Water, Compound	160
Lees	172
Lees Ashes	178
Lime Water	191
Lob	11
Lovage Cordial	150
Luting	13
Maceration	200, 208
Malt	211
Malt Distilling	91
Malt Whisky	96
Malting	63
Mashing	67
Mashing of Potatoes	114
Materials, Raw	214
Method, English	81
Method, French	79
Mode of Operating	32, 39
Molasses, Beet-Root	170
Molasses, Distillation of	140
Musk, Tincture of	197
Nectar, Process for Making	149
Noyau, French	151
Oats for Distilling	64
Observations on Special Distillations	165
Oil, Empyreumatic	173
Operating, Mode of	32
Orange-flower Water	158
Orange Wine	159
Peach Brandy	136
Peppermint Cordial	152
Peppermint Water	158
Piquette	172
Potatoes, Mashing of	114
Potatoes, Reduction of	112
Potatoes, Spirit of	106
Preparation of Cordials, etc.	187
Preparing Brandy	93
Preservation of Spirituous Liquors	213
Prevention of Deterioration of Brandies	95

INDEX.

	PAGE
Prevention of Inequality of Heat.	58
Process of Malting	63
Putrid Fermentation	202
Raisin Spirits	143
Rasp, Beet	128
Raspberry Brandy	147
Rasping Potatoes	116
Ratafia, Imperial	149
Raw Materials	214
Receiver	45, 47
Recovering	14
Rectification	29, 89, 205, 209
Rectification into Hollands Gin	103
Rectifier	28
Reduction of Potatoes	112
Refrigerator	31
Refrigeratory	47
Regulator	32
Repasses	24
Reservoir	32
Retorts	45
Rice, Spirits of	124
Rosemary Water	157
Rules for Determining the Relative Value and Strength of Spirits	164
Rum	137
Rum Shrub, Process for Making	144
Rye for Distilling	64
Saccharification	68
Sand Bath	48
Season for Malting	74
Selection of Apparatus	165
Separation of Fecula	116
Shrub, Brandy	145
Shrub, Rum	144
Simple Lavender Water	159
Simple Waters, Distillation of	155
Special Distillations	165
Spirit of Potatoes	106
Spirits	12
Spirits, Flavoring and Coloring of	143
Spirits of Beet-Root	127
Spirits of Cherries	133
Spirits of Corn	63
Spirits of Raisins	143
Spirits of Rice	124

	PAGE
Spirits, Rules for Strength and Value	164
Spirituous Liquors, Preservation of	213
Spirituous Waters	161
Steeping	65
Still, Charge of	12
Stills	17, 27, 51, 52
Stills for Simple Waters	156
Stomachic Compound	191
Strength of Spirits, Rules for	164
Sulphuric Ether	191
Tampot	18
Tantern	23
Testing Wines	184
Tinctures	194
Tincture of Musk	197
Tonic and Alterative Cordial	195
Uses of Feints	163
Usquebaugh, Irish	148
Utensils for Distilling	208
Value of Spirits, Rules for	164
Vinegar, English	154
Vinegar, French	153
Vinous Fermentation	202
Water, Cinnamon	158
Water, Compound Lavender	160
Water, Damask-Rose	158
Water, Hungary	160
Water, Jessamine	162
Water, Lavender	159
Water, Lime	191
Water of Cherries	133
Water, Orange-flower	158
Water, Peppermint	158
Water, Rosemary	157
Waters, Simple	155
Waters, Spirituous	161
Wheat for Distilling	64
Whisky, Malt	96
Wine, Orange	159
Wine-warming Condenser	30
Wines for Distillation	93
Wines, Testing of	184
Worts, Gravity of	11
Yeast	210

CATALOGUE

OF

PRACTICAL AND SCIENTIFIC BOOKS,

PUBLISHED BY

HENRY CAREY BAIRD & CO.,

Industrial Publishers and Booksellers,

NO. 810 WALNUT STREET,

PHILADELPHIA.

☞ Any of the Books comprised in this Catalogue will be sent by mail, free of postage, at the publication price.

☞ A Descriptive Catalogue, 96 pages, 8vo., will be sent, free of postage, to any one who will furnish the publisher with his address.

ARLOT.—A Complete Guide for Coach Painters.
Translated from the French of M. ARLOT, Coach Painter; for eleven years Foreman of Painting to M. Eherler, Coach Maker, Paris. By A. A. FESQUET, Chemist and Engineer. To which is added an Appendix, containing Information respecting the Materials and the Practice of Coach and Car Painting and Varnishing in the United States and Great Britain. 12mo. $1.25

ARMENGAUD, AMOROUX, and JOHNSON.—The Practical Draughtsman's Book of Industrial Design, and Machinist's and Engineer's Drawing Companion:
Forming a Complete Course of Mechanical Engineering and Architectural Drawing. From the French of M. Armengaud the elder, Prof. of Design in the Conservatoire of Arts and Industry, Paris, and MM. Armengaud the younger, and Amoroux, Civil Engineers. Rewritten and arranged with additional matter and plates, selections from and examples of the most useful and generally employed mechanism of the day. By WILLIAM JOHNSON, Assoc. Inst. C. E., Editor of "The Practical Mechanic's Journal." Illustrated by 50 folio steel plates, and 50 wood-cuts. A new edition, 4to. $10.00

ARROWSMITH.—Paper-Hanger's Companion:
A Treatise in which the Practical Operations of the Trade are Systematically laid down: with Copious Directions Preparatory to Papering; Preventives against the Effect of Damp on Walls; the Various Cements and Pastes Adapted to the Several Purposes of the Trade; Observations and Directions for the Panelling and Ornamenting of Rooms, etc. By JAMES ARROWSMITH, Author of "Analysis of Drapery," etc. 12mo., cloth. $1.25

ASHTON.—The Theory and Practice of the Art of Designing Fancy Cotton and Woollen Cloths from Sample:
Giving full Instructions for Reducing Drafts, as well as the Methods of Spooling and Making out Harness for Cross Drafts, and Finding any Required Reed, with Calculations and Tables of Yarn. By FREDERICK T. ASHTON, Designer, West Pittsfield, Mass. With 52 Illustrations. One volume, 4to. $10.00

BAIRD.—Letters on the Crisis, the Currency and the Credit System.
By HENRY CAREY BAIRD. Pamphlet. 05

BAIRD.—Protection of Home Labor and Home Productions necessary to the Prosperity of the American Farmer.
By HENRY CAREY BAIRD. 8vo., paper. 10

BAIRD.—Some of the Fallacies of British Free-Trade Revenue Reform.
Two Letters to Arthur Latham Perry, Professor of History and Political Economy in Williams College. By HENRY CAREY BAIRD. Pamphlet. 05

BAIRD.—The Rights of American Producers, and the Wrongs of British Free-Trade Revenue Reform.
By HENRY CAREY BAIRD. Pamphlet. 05

BAIRD.—Standard Wages Computing Tables:
An Improvement in all former Methods of Computation, so arranged that wages for days, hours, or fractions of hours, at a specified rate per day or hour, may be ascertained at a glance. By T. SPANGLER BAIRD. Oblong folio. $5.00

BAIRD.—The American Cotton Spinner, and Manager's and Carder's Guide:
A Practical Treatise on Cotton Spinning; giving the Dimensions and Speed of Machinery, Draught and Twist Calculations, etc.; with notices of recent Improvements: together with Rules and Examples for making changes in the sizes and numbers of Roving and Yarn. Compiled from the papers of the late ROBERT H. BAIRD. 12mo. $1.50

BAKER.—Long-Span Railway Bridges:
Comprising Investigations of the Comparative Theoretical and Practical Advantages of the various Adopted or Proposed Type Systems of Construction; with numerous Formulæ and Tables. By B. BAKER. 12mo. $2.00

BAUERMAN.—A Treatise on the Metallurgy of Iron:
Containing Outlines of the History of Iron Manufacture, Methods of Assay, and Analysis of Iron Ores, Processes of Manufacture of Iron and Steel, etc., etc. By H. BAUERMAN, F. G. S., Associate of the Royal School of Mines. First American Edition, Revised and Enlarged. With an Appendix on the Martin Process for Making Steel, from the Report of ABRAM S. HEWITT, U. S. Commissioner to the Universal Exposition at Paris, 1867. Illustrated. 12mo. . $2.00

BEANS.—A Treatise on Railway Curves and the Location of Railways.
By E. W. BEANS, C. E. Illustrated. 12mo. Tucks. . . $1.50

BELL.—Carpentry Made Easy:
Or, The Science and Art of Framing on a New and Improved System. With Specific Instructions for Building Balloon Frames, Barn Frames, Mill Frames, Warehouses, Church Spires, etc. Comprising also a System of Bridge Building, with Bills, Estimates of Cost, and valuable Tables. Illustrated by 38 plates, comprising nearly 200 figures. By WILLIAM E. BELL, Architect and Practical Builder. 8vo. . $5.00

BELL.—Chemical Phenomena of Iron Smelting:
An Experimental and Practical Examination of the Circumstances which determine the Capacity of the Blast Furnace, the Temperature of the Air, and the proper Condition of the Materials to be operated upon. By I. LOWTHIAN BELL. Illustrated. 8vo. . . $6.00

BEMROSE.—Manual of Wood Carving:
With Practical Illustrations for Learners of the Art, and Original and Selected Designs. By WILLIAM BEMROSE, Jr. With an Introduction by LLEWELLYN JEWITT, F. S. A., etc. With 128 Illustrations. 4to., cloth. $3.00

BICKNELL.—Village Builder, and Supplement:
Elevations and Plans for Cottages, Villas, Suburban Residences, Farm Houses, Stables and Carriage Houses, Store Fronts, School Houses, Churches, Court Houses, and a model Jail; also, Exterior and Interior details for Public and Private Buildings, with approved Forms of Contracts and Specifications, including Prices of Building Materials and Labor at Boston, Mass., and St. Louis, Mo. Containing 75 plates drawn to scale; showing the style and cost of building in different sections of the country, being an original work comprising the designs of twenty leading architects, representing the New England, Middle, Western, and Southwestern States. 4to. . $10.00

BLENKARN.—Practical Specifications of Works executed in Architecture, Civil and Mechanical Engineering, and in Road Making and Sewering:
To which are added a series of practically useful Agreements and Reports. By JOHN BLENKARN. Illustrated by 15 large folding plates. 8vo. $9.00

BLINN.—A Practical Workshop Companion for Tin, Sheet-Iron, and Copperplate Workers:
Containing Rules for describing various kinds of Patterns used by Tin, Sheet-Iron, and Copper-plate Workers; Practical Geometry; Mensuration of Surfaces and Solids; Tables of the Weights of Metals, Lead Pipe, etc.; Tables of Areas and Circumferences of Circles; Japan, Varnishes, Lackers, Cements, Compositions, etc., etc. By LEROY J. BLINN, Master Mechanic. With over 100 Illustrations. 12mo. $2.50

BOOTH.—Marble Worker's Manual:
Containing Practical Information respecting Marbles in general, their Cutting, Working, and Polishing; Veneering of Marble; Mosaics; Composition and Use of Artificial Marble, Stuccos, Cements, Receipts, Secrets, etc., etc. Translated from the French by M. L. BOOTH. With an Appendix concerning American Marbles. 12mo., cloth. $1.50

BOOTH AND MORFIT.—The Encyclopedia of Chemistry, Practical and Theoretical:
Embracing its application to the Arts, Metallurgy, Mineralogy, Geology, Medicine, and Pharmacy. By JAMES C. BOOTH, Melter and Refiner in the United States Mint, Professor of Applied Chemistry in the Franklin Institute, etc., assisted by CAMPBELL MORFIT, author of "Chemical Manipulations," etc. Seventh edition. Royal 8vo., 978 pages, with numerous wood-cuts and other illustrations. . $5.00

BOX.—A Practical Treatise on Heat:
As applied to the Useful Arts; for the Use of Engineers, Architects, etc. By THOMAS BOX, author of "Practical Hydraulics." Illustrated by 14 plates containing 114 figures. 12mo. $5.00

BOX.—Practical Hydraulics:
A Series of Rules and Tables for the use of Engineers, etc. By THOMAS BOX. 12mo. $2.50

BROWN.—Five Hundred and Seven Mechanical Movements:
Embracing all those which are most important in Dynamics, Hydraulics, Hydrostatics, Pneumatics, Steam Engines, Mill and other Gearing, Presses, Horology, and Miscellaneous Machinery; and including many movements never before published, and several of which have only recently come into use. By HENRY T. BROWN, Editor of the "American Artisan." In one volume, 12mo. . . . $1.00

BUCKMASTER.—The Elements of Mechanical Physics:
By J. C. BUCKMASTER, late Student in the Government School of Mines; Certified Teacher of Science by the Department of Science and Art; Examiner in Chemistry and Physics in the Royal College of Preceptors; and late Lecturer in Chemistry and Physics of the Royal Polytechnic Institute. Illustrated with numerous engravings. In one volume, 12mo. $1.50

BULLOCK.—The American Cottage Builder:
A Series of Designs, Plans, and Specifications, from $200 to $20,000, for Homes for the People; together with Warming, Ventilation, Drainage, Painting, and Landscape Gardening. By JOHN BULLOCK, Architect, Civil Engineer, Mechanician, and Editor of "The Rudiments of Architecture and Building," etc., etc. Illustrated by 75 engravings. In one volume, 8vo. $3.50

BULLOCK.—The Rudiments of Architecture and Building:
For the use of Architects, Builders, Draughtsmen, Machinists, Engineers, and Mechanics. Edited by JOHN BULLOCK, author of "The American Cottage Builder." Illustrated by 250 engravings. In one volume, 8vo. $3.50

BURGH.—Practical Illustrations of Land and Marine Engines:
Showing in detail the Modern Improvements of High and Low Pressure, Surface Condensation, and Super-heating, together with Land and Marine Boilers. By N. P. BURGH, Engineer. Illustrated by 20 plates, double elephant folio, with text. . . . $21.00

BURGH.—Practical Rules for the Proportions of Modern Engines and Boilers for Land and Marine Purposes.
By N. P. BURGH, Engineer. 12mo. $1.50

BURGH.—The Slide-Valve Practically Considered.
By N. P. BURGH, Engineer. Completely illustrated. 12mo. $2.00

BYLES.—Sophisms of Free Trade and Popular Political Economy Examined.
By a BARRISTER (Sir JOHN BARNARD BYLES, Judge of Common Pleas). First American from the Ninth English Edition, as published by the Manchester Reciprocity Association. In one volume, 12mo. $1.25

BYRN.—The Complete Practical Brewer:
Or Plain, Accurate, and Thorough Instructions in the Art of Brewing Beer, Ale, Porter, including the Process of making Bavarian Beer, all the Small Beers, such as Root-beer, Ginger-pop, Sarsaparilla-beer, Mead, Spruce Beer, etc., etc. Adapted to the use of Public Brewers and Private Families. By M. LA FAYETTE BYRN, M D. With illustrations. 12mo. $1.25

BYRN.—The Complete Practical Distiller:
Comprising the most perfect and exact Theoretical and Practical Description of the Art of Distillation and Rectification; including all of the most recent improvements in distilling apparatus; instructions for preparing spirits from the numerous vegetables, fruits, etc.; directions for the distillation and preparation of all kinds of brandies and other spirits, spirituous and other compounds, etc., etc. By M. LA FAYETTE BYRN, M. D. Eighth Edition. To which are added, Practical Directions for Distilling, from the French of Th. Fling, Brewer and Distiller. 12mo. $1.50

BYRNE.—Handbook for the Artisan, Mechanic, and Engineer:
Comprising the Grinding and Sharpening of Cutting Tools, Abrasive Processes, Lapidary Work, Gem and Glass Engraving, Varnishing and Lackering, Apparatus, Materials and Processes for Grinding and Polishing, etc. By OLIVER BYRNE. Illustrated by 185 wood engravings. In one volume, 8vo. $5.00

BYRNE.—Pocket Book for Railroad and Civil Engineers:
Containing New, Exact, and Concise Methods for Laying out Railroad Curves, Switches, Frog Angles, and Crossings; the Staking out of work; Levelling; the Calculation of Cuttings; Embankments; Earth-work, etc. By OLIVER BYRNE. 18mo., full bound, pocketbook form. $1.75

BYRNE.—The Practical Model Calculator:
For the Engineer, Mechanic, Manufacturer of Engine Work, Naval Architect, Miner, and Millwright. By OLIVER BYRNE. 1 volume, 8vo., nearly 600 pages $4.50

BYRNE.—The Practical Metal-Worker's Assistant:
Comprising Metallurgic Chemistry; the Arts of Working all Metals and Alloys; Forging of Iron and Steel; Hardening and Tempering; Melting and Mixing; Casting and Founding; Works in Sheet Metal; The Processes Dependent on the Ductility of the Metals; Soldering; and the most Improved Processes and Tools employed by Metal-Workers. With the Application of the Art of Electro-Metallurgy to Manufacturing Processes; collected from Original Sources, and from the Works of Holtzapffel, Bergeron, Leupold, Plumier, Napier, Scoffern, Clay, Fairbairn, and others. By OLIVER BYRNE. A new, revised, and improved edition, to which is added An Appendix, containing THE MANUFACTURE OF RUSSIAN SHEET-IRON. By JOHN PERCY, M. D., F.R.S. THE MANUFACTURE OF MALLEABLE IRON CASTINGS, and IMPROVEMENTS IN BESSEMER STEEL. By A. A. FESQUET, Chemist and Engineer. With over 600 Engravings, illustrating every Branch of the Subject. 8vo. . . . $7.00

Cabinet Maker's Album of Furniture:
Comprising a Collection of Designs for Furniture. Illustrated by 48 Large and Beautifully Engraved Plates. In one vol., oblong $3.50

CALLINGHAM.—Sign Writing and Glass Embossing:
A Complete Practical Illustrated Manual of the Art. By JAMES CALLINGHAM. In one volume, 12mo. $1.50

CAMPIN.—A Practical Treatise on Mechanical Engineering:
Comprising Metallurgy, Moulding, Casting, Forging, Tools, Workshop Machinery, Mechanical Manipulation, Manufacture of Steam-engines, etc., etc. With an Appendix on the Analysis of Iron and Iron Ores. By FRANCIS CAMPIN, C. E. To which are added, Observations on the Construction of Steam Boilers, and Remarks upon Furnaces used for Smoke Prevention; with a Chapter on Explosions. By R. Armstrong, C. E., and John Bourne. Rules for Calculating the Change Wheels for Screws on a Turning Lathe, and for a Wheel-cutting Machine. By J. LA NICCA. Management of Steel, Including Forging, Hardening, Tempering, Annealing, Shrinking, and Expansion. And the Case-hardening of Iron. By G. EDE. 8vo. Illustrated with 29 plates and 100 wood engravings . . . $6.00

CAMPIN.—The Practice of Hand-Turning in Wood, Ivory, Shell, etc.:
With Instructions for Turning such works in Metal as may be required in the Practice of Turning Wood, Ivory, etc. Also, an Appendix on Ornamental Turning. By FRANCIS CAMPIN; with Numerous Illustrations. 12mo., cloth $3.00

CAREY.—The Works of Henry C. Carey:
FINANCIAL CRISES, their Causes and Effects. 8vo. paper . 25
HARMONY OF INTERESTS: Agricultural, Manufacturing, and Commercial. 8vo., cloth $1.50
MANUAL OF SOCIAL SCIENCE. Condensed from Carey's "Principles of Social Science." By KATE MCKEAN. 1 vol. 12mo. $2.25
MISCELLANEOUS WORKS: comprising "Harmony of Interests," "Money," "Letters to the President," "Financial Crises," "The Way to Outdo England Without Fighting Her," "Resources of the Union," "The Public Debt," "Contraction or Expansion?" "Review of the Decade 1857–'67," "Reconstruction," etc., etc. Two vols., 8vo., cloth . . . .
PAST, PRESENT, AND FUTURE. 8vo. $2.50
PRINCIPLES OF SOCIAL SCIENCE. 3 vols., 8vo., cloth $10.00
THE SLAVE-TRADE, DOMESTIC AND FOREIGN; Why it Exists, and How it may be Extinguished (1853). 8vo., cloth . $2.00
LETTERS ON INTERNATIONAL COPYRIGHT (1867) . 50
THE UNITY OF LAW: As Exhibited in the Relations of Physical, Social, Mental, and Moral Science (1872). In one volume, 8vo., pp. xxiii., 433. Cloth $3.50

CHAPMAN.—A Treatise on Ropemaking:
As Practised in private and public Rope yards, with a Description of the Manufacture, Rules, Tables of Weights, etc., adapted to the Trades, Shipping, Mining, Railways, Builders, etc. By ROBERT CHAPMAN. 24mo. $1.50

COLBURN.—The Locomotive Engine:
Including a Description of its Structure, Rules for Estimating its Capabilities, and Practical Observations on its Construction and Management. By ZERAH COLBURN. Illustrated. A new edition. 12mo. $1.25

CRAIK.—The Practical American Millwright and Miller.
By DAVID CRAIK, Millwright. Illustrated by numerous wood engravings, and two folding plates. 8vo. $5.00

DE GRAFF.—The Geometrical Stair Builders' Guide:
Being a Plain Practical System of Hand-Railing, embracing all its necessary Details, and Geometrically Illustrated by 22 Steel Engravings; together with the use of the most approved principles of Practical Geometry. By SIMON DE GRAFF, Architect. 4to. . $5.00

DE KONINCK.—DIETZ.—A Practical Manual of Chemical Analysis and Assaying:
As applied to the Manufacture of Iron from its Ores, and to Cast Iron, Wrought Iron, and Steel, as found in Commerce. By L. L. DE KONINCK, Dr. Sc., and E. DIETZ, Engineer. Edited with Notes, by ROBERT MALLET, F.R.S., F.S.G., M.I.C.E., etc. American Edition, Edited with Notes and an Appendix on Iron Ores, by A. A. FESQUET, Chemist and Engineer. One volume, 12mo. $2.50

DUNCAN.—Practical Surveyor's Guide:
Containing the necessary information to make any person, of common capacity, a finished land surveyor without the aid of a teacher. By ANDREW DUNCAN. Illustrated. 12mo., cloth. . . . $1.25

DUPLAIS.—A Treatise on the Manufacture and Distillation of Alcoholic Liquors:
Comprising Accurate and Complete Details in Regard to Alcohol from Wine, Molasses, Beets, Grain, Rice, Potatoes, Sorghum, Asphodel, Fruits, etc.; with the Distillation and Rectification of Brandy, Whiskey, Rum, Gin, Swiss Absinthe, etc., the Preparation of Aromatic Waters, Volatile Oils or Essences, Sugars, Syrups, Aromatic Tinctures, Liqueurs, Cordial Wines, Effervescing Wines, etc., the Aging of Brandy and the Improvement of Spirits, with Copious Directions and Tables for Testing and Reducing Spirituous Liquors, etc., etc. Translated and Edited from the French of MM. DUPLAIS, Ainé et Jeune. By M. McKENNIE, M.D. To which are added the United States Internal Revenue Regulations for the Assessment and Collection of Taxes on Distilled Spirits. Illustrated by fourteen folding plates and several wood engravings. 743 pp., 8vo. $10.00

DUSSAUCE.—A General Treatise on the Manufacture of Every Description of Soap:
Comprising the Chemistry of the Art, with Remarks on Alkalies, Saponifiable Fatty Bodies, the apparatus necessary in a Soap Factory, Practical Instructions in the manufacture of the various kinds of Soap, the assay of Soaps, etc., etc. Edited from Notes of Larmé, Fontenelle, Malapayre, Dufour, and others, with large and important additions by Prof. H. DUSSAUCE, Chemist. Illustrated. In one vol., 8vo. . $17.50

DUSSAUCE.—A General Treatise on the Manufacture of Vinegar:
Theoretical and Practical. Comprising the various Methods, by the Slow and the Quick Processes, with Alcohol, Wine, Grain, Malt, Cider, Molasses, and Beets; as well as the Fabrication of Wood Vinegar, etc., etc. By Prof. H. DUSSAUCE. In one volume, 8vo. . . $5.00

DUSSAUCE.—A New and Complete Treatise on the Arts of Tanning, Currying, and Leather Dressing:
Comprising all the Discoveries and Improvements made in France, Great Britain, and the United States. Edited from Notes and Documents of Messrs. Sallerou, Grouvelle, Duval, Dessables, Labarraque, Payen, René, De Fontenelle, Malapeyre, etc., etc. By Prof. H. DUSSAUCE, Chemist. Illustrated by 212 wood engravings. 8vo. $25.00

DUSSAUCE.—A Practical Guide for the Perfumer:
Being a New Treatise on Perfumery, the most favorable to the Beauty without being injurious to the Health, comprising a Description of the substances used in Perfumery, the Formulæ of more than 1000 Preparations; such as Cosmetics, Perfumed Oils, Tooth Powders, Waters, Extracts, Tinctures, Infusions, Spirits, Vinaigres, Essential Oils, Pastels, Creams, Soaps, and many new Hygienic Products not hitherto described. Edited from Notes and Documents of Messrs. Debay, Lanel, etc. With additions by Prof. H. DUSSAUCE, Chemist. 12mo.

DUSSAUCE.—Practical Treatise on the Fabrication of Matches, Gun Cotton, and Fulminating Powders.
By Prof. H. DUSSAUCE. 12mo. $3.00

Dyer and Color-maker's Companion:
Containing upwards of 200 Receipts for making Colors, on the most approved principles, for all the various styles and fabrics now in existence; with the Scouring Process, and plain Directions for Preparing, Washing-off, and Finishing the Goods. In one vol., 12mo. . $1.25

EASTON.—A Practical Treatise on Street or Horse-power Railways.
By ALEXANDER EASTON, C. E. Illustrated by 23 plates. 8vo., cloth. $3.00

ELDER.—Questions of the Day:
Economic and Social. By Dr. WILLIAM ELDER. 8vo. . $3.00

FAIRBAIRN.—The Principles of Mechanism and Machinery of Transmission:
Comprising the Principles of Mechanism, Wheels, and Pulleys, Strength and Proportions of Shafts, Coupling of Shafts, and Engaging and Disengaging Gear. By Sir WILLIAM FAIRBAIRN, C.E., LL.D., F.R.S., F.G.S. Beautifully illustrated by over 150 wood-cuts. In one volume, 12mo. $2.50

FORSYTH.—Book of Designs for Headstones, Mural, and other Monuments:
Containing 78 Designs. By JAMES FORSYTH. With an Introduction by CHARLES BOUTELL, M. A. 4to., cloth. $5.00

GIBSON.—The American Dyer:
A Practical Treatise on the Coloring of Wool, Cotton, Yarn and Cloth, in three parts. Part First gives a descriptive account of the Dye Stuffs; if of vegetable origin, where produced, how cultivated, and how prepared for use; if chemical, their composition, specific gravities, and general adaptability, how adulterated, and how to detect the adulterations, etc. Part Second is devoted to the Coloring of Wool, giving recipes for one hundred and twenty-nine different colors or shades, and is supplied with sixty colored samples of Wool. Part Third is devoted to the Coloring of Raw Cotton or Cotton Waste, for mixing with Wool Colors in the Manufacture of all kinds of Fabrics, gives recipes for thirty-eight different colors or shades, and is supplied with twenty-four colored samples of Cotton Waste. Also, recipes for Coloring Beavers, Doeskins, and Flannels, with remarks upon Anilines, giving recipes for fifteen different colors or shades, and nine samples of Aniline Colors that will stand both the Fulling and Scouring process. Also, recipes for Aniline Colors on Cotton Thread, and recipes for Common Colors on Cotton Yarns. Embracing in all over two hundred recipes for Colors and Shades, and ninety-four samples of Colored Wool and Cotton Waste, etc. By RICHARD H. GIBSON, Practical Dyer and Chemist. In one volume, 8vo. . . $6.00

GILBART.—History and Principles of Banking:
A Practical Treatise. By JAMES W. GILBART, late Manager of the London and Westminster Bank. With additions. In one volume, 8vo., 600 pages, sheep $5.00

Gothic Album for Cabinet Makers:
Comprising a Collection of Designs for Gothic Furniture. Illustrated by 23 large and beautifully engraved plates. Oblong . . $2.00

GRANT.—Beet-root Sugar and Cultivation of the Beet.
By E. B. GRANT. 12mo. $1.25

GREGORY.—Mathematics for Practical Men:
Adapted to the Pursuits of Surveyors, Architects, Mechanics, and Civil Engineers. By OLINTHUS GREGORY. 8vo., plates, cloth $3.00

GRISWOLD.—Railroad Engineer's Pocket Companion for the Field:
Comprising Rules for Calculating Deflection Distances and Angles, Tangential Distances and Angles, and all Necessary Tables for Engineers; also the art of Levelling from Preliminary Survey to the Construction of Railroads, intended Expressly for the Young Engineer, together with Numerous Valuable Rules and Examples. By W. GRISWOLD. 12mo., tucks $1.75

GRUNER.—Studies of Blast Furnace Phenomena.
By M. L. GRUNER, President of the General Council of Mines of France, and lately Professor of Metallurgy at the Ecole des Mines. Translated, with the Author's sanction, with an Appendix, by L. D. B. Gordon, F. R. S. E.. F. G. S. Illustrated. 8vo. . . . $2.50

GUETTIER.—Metallic Alloys:

Being a Practical Guide to their Chemical and Physical Properties, their Preparation, Composition, and Uses. Translated from the French of A. GUETTIER, Engineer and Director of Foundries, author of "La Fouderie en France," etc., etc. By A. A. FESQUET, Chemist and Engineer. In one volume, 12mo. $3.00

HARRIS.—Gas Superintendent's Pocket Companion.

By HARRIS & BROTHER, Gas Meter Manufacturers, 1115 and 1117 Cherry Street, Philadelphia. Full bound in pocket-book form $1.00

Hats and Felting:

A Practical Treatise on their Manufacture. By a Practical Hatter. Illustrated by Drawings of Machinery, etc. 8vo. . . . $1.25

HOFMANN.—A Practical Treatise on the Manufacture of Paper in all its Branches.

By CARL HOFMANN. Late Superintendent of paper mills in Germany and the United States; recently manager of the Public Ledger Paper Mills, near Elkton, Md. Illustrated by 110 wood engravings, and five large folding plates. In one volume, 4to., cloth; 398 pages $15.00

HUGHES.—American Miller and Millwright's Assistant.

By WM. CARTER HUGHES. A new edition. In one vol., 12mo. $1.50

HURST.—A Hand-Book for Architectural Surveyors and others engaged in Building:

Containing Formulæ useful in Designing Builder's work, Table of Weights, of the materials used in Building, Memoranda connected with Builders' work, Mensuration, the Practice of Builders' Measurement, Contracts of Labor, Valuation of Property, Summary of the Practice in Dilapidation, etc., etc. By J. F. HURST, C. E. Second edition, pocket-book form, full bound $2.00

JERVIS.—Railway Property:

A Treatise on the Construction and Management of Railways; designed to afford useful knowledge, in the popular style, to the holders of this class of property; as well as Railway Managers, Officers, and Agents. By JOHN B. JERVIS, late Chief Engineer of the Hudson River Railroad, Croton Aqueduct, etc. In one vol., 12mo., cloth $2.00

JOHNSTON.—Instructions for the Analysis of Soils, Limestones, and Manures.

By J. F. W. JOHNSTON. 12mo. . . . . . .

KEENE.—A Hand-Book of Practical Gauging:
For the Use of Beginners, to which is added, A Chapter on Distillation, describing the process in operation at the Custom House for ascertaining the strength of wines. By JAMES B. KEENE, of H. M. Customs. 8vo. $1.25

KELLEY.—Speeches, Addresses, and Letters on Industrial and Financial Questions.
By Hon. WILLIAM D. KELLEY, M. C. In one volume, 544 pages, 8vo. $3.00

KENTISH.—A Treatise on a Box of Instruments,
And the Slide Rule; with the Theory of Trigonometry and Logarithms, including Practical Geometry, Surveying, Measuring of Timber, Cask and Malt Gauging, Heights, and Distances. By THOMAS KENTISH. In one volume. 12mo. $1.25

KOBELL.—ERNI.—Mineralogy Simplified:
A short Method of Determining and Classifying Minerals, by means of simple Chemical Experiments in the Wet Way. Translated from the last German Edition of F. VON KOBELL, with an Introduction to Blow-pipe Analysis and other additions. By HENRI ERNI, M. D., late Chief Chemist, Department of Agriculture, author of "Coal Oil and Petroleum." In one volume, 12mo. $2.50

LANDRIN.—A Treatise on Steel:
Comprising its Theory, Metallurgy, Properties, Practical Working, and Use. By M. H. C. LANDRIN, Jr., Civil Engineer. Translated from the French, with Notes, by A. A. FESQUET, Chemist and Engineer. With an Appendix on the Bessemer and the Martin Processes for Manufacturing Steel, from the Report of Abram S. Hewitt, United States Commissioner to the Universal Exposition, Paris, 1867. In one volume, 12mo. $3.00

LARKIN.—The Practical Brass and Iron Founder's Guide:
A Concise Treatise on Brass Founding, Moulding, the Metals and their Alloys, etc.: to which are added Recent Improvements in the Manufacture of Iron, Steel by the Bessemer Process, etc., etc. By JAMES LARKIN, late Conductor of the Brass Foundry Department in Reany, Neafie & Co's. Penn Works, Philadelphia. Fifth edition, revised, with Extensive additions. In one volume, 12mo. . . $2.25

LEAVITT.—Facts about Peat as an Article of Fuel:
With Remarks upon its Origin and Composition, the Localities in which it is found, the Methods of Preparation and Manufacture, and the various Uses to which it is applicable; together with many other matters of Practical and Scientific Interest. To which is added a chapter on the Utilization of Coal Dust with Peat for the Production of an Excellent Fuel at Moderate Cost, specially adapted for Steam Service. By T. H. LEAVITT. Third edition. 12mo. . . . $1.75

LEROUX, C.—A Practical Treatise on the Manufacture of Worsteds and Carded Yarns:
Comprising Practical Mechanics, with Rules and Calculations applied to Spinning; Sorting, Cleaning, and Scouring Wools; the English and French methods of Combing, Drawing, and Spinning Worsteds and Manufacturing Carded Yarns. Translated from the French of CHARLES LEROUX, Mechanical Engineer, and Superintendent of a Spinning Mill, by HORATIO PAINE, M. D., and A. A. FESQUET, Chemist and Engineer. Illustrated by 12 large Plates. To which is added an Appendix, containing extracts from the Reports of the International Jury, and of the Artisans selected by the Committee appointed by the Council of the Society of Arts, London, on Woollen and Worsted Machinery and Fabrics, as exhibited in the Paris Universal Exposition, 1867. 8vo., cloth. $5.00

LESLIE (Miss).—Complete Cookery:
Directions for Cookery in its Various Branches. By MISS LESLIE. 60th thousand. Thoroughly revised, with the addition of New Receipts. In one volume, 12mo., cloth. $1.50

LESLIE (Miss).—Ladies' House Book:
A Manual of Domestic Economy. 20th revised edition. 12mo., cloth.

LESLIE (Miss).—Two Hundred Receipts in French Cookery.
Cloth, 12mo.

LIEBER.—Assayer's Guide:
Or, Practical Directions to Assayers, Miners, and Smelters, for the Tests and Assays, by Heat and by Wet Processes, for the Ores of all the principal Metals, of Gold and Silver Coins and Alloys, and of Coal, etc. By OSCAR M. LIEBER. 12mo., cloth. . . $1.25

LOTH.—The Practical Stair Builder:
A Complete Treatise on the Art of Building Stairs and Hand-Rails, Designed for Carpenters, Builders, and Stair-Builders. Illustrated with Thirty Original Plates. By C. EDWARD LOTH, Professional Stair-Builder. One large 4to. volume. $10.00

LOVE.—The Art of Dyeing, Cleaning, Scouring, and Finishing, on the Most Approved English and French Methods:
Being Practical Instructions in Dyeing Silks, Woollens, and Cottons, Feathers, Chips, Straw, etc. Scouring and Cleaning Bed and Window Curtains, Carpets, Rugs, etc. French and English Cleaning, any Color or Fabric of Silk, Satin, or Damask. By THOMAS LOVE, a Working Dyer and Scourer. Second American Edition, to which are added General Instructions for the Use of Aniline Colors. In one volume, 8vo., 343 pages. $5.00

MAIN and BROWN.—Questions on Subjects Connected with the Marine Steam-Engine:

And Examination Papers: with Hints for their Solution. By THOMAS J. MAIN, Professor of Mathematics, Royal Naval College, and THOMAS BROWN, Chief Engineer, R. N. 12mo., cloth. . . . $1.50

MAIN and BROWN.—The Indicator and Dynamometer:

With their Practical Applications to the Steam-Engine. By THOMAS J. MAIN, M. A. F. R., Assistant Professor Royal Naval College, Portsmouth, and THOMAS BROWN, Assoc. Inst. C. E., Chief Engineer, R. N., attached to the Royal Naval College. Illustrated. From the Fourth London Edition. 8vo. $1.50

MAIN and BROWN.—The Marine Steam-Engine.

By THOMAS J. MAIN, F. R.; Assistant S. Mathematical Professor at the Royal Naval College, Portsmouth, and THOMAS BROWN, Assoc. Inst. C. E., Chief Engineer R. N. Attached to the Royal Naval College. Authors of "Questions connected with the Marine Steam-Engine," and the "Indicator and Dynamometer." With numerous Illustrations. In one volume, 8vo. $5.00

MARTIN.—Screw-Cutting Tables, for the Use of Mechanical Engineers:

Showing the Proper Arrangement of Wheels for Cutting the Threads of Screws of any required Pitch; with a Table for Making the Universal Gas-Pipe Thread and Taps. By W. A. MARTIN, Engineer. 8vo. 50

Mechanics' (Amateur) Workshop:

A treatise containing plain and concise directions for the manipulation of Wood and Metals, including Casting, Forging, Brazing, Soldering, and Carpentry. By the author of the "Lathe and its Uses." Third edition. Illustrated. 8vo. $3.00

MOLESWORTH.—Pocket-Book of Useful Formulæ and Memoranda for Civil and Mechanical Engineers.

By GUILFORD L. MOLESWORTH, Member of the Institution of Civil Engineers, Chief Resident Engineer of the Ceylon Railway. Second American, from the Tenth London Edition. In one volume, full bound in pocket-book form. $2.00

NAPIER.—A System of Chemistry Applied to Dyeing.

By JAMES NAPIER, F. C. S. A New and Thoroughly Revised Edition. Completely brought up to the present state of the Science, including the Chemistry of Coal Tar Colors, by A. A. FESQUET, Chemist and Engineer. With an Appendix on Dyeing and Calico Printing, as shown at the Universal Exposition, Paris, 1867. Illustrated. In one volume, 8vo., 422 pages. $5.00

NAPIER.—Manual of Electro-Metallurgy:
Including the Application of the Art to Manufacturing Processes. By JAMES NAPIER. Fourth American, from the Fourth London edition, revised and enlarged. Illustrated by engravings. In one vol., 8vo. $2.00

NASON.—Table of Reactions for Qualitative Chemical Analysis:
By HENRY B. NASON, Professor of Chemistry in the Rensselaer Polytechnic Institute, Troy, New York. Illustrated by Colors. . 63

NEWBERY.—Gleanings from Ornamental Art of every style:
Drawn from Examples in the British, South Kensington, Indian, Crystal Palace, and other Museums, the Exhibitions of 1851 and 1862, and the best English and Foreign works. In a series of one hundred exquisitely drawn Plates, containing many hundred examples. By ROBERT NEWBERY. 4to. $12.50

NICHOLSON.—A Manual of the Art of Bookbinding:
Containing full instructions in the different Branches of Forwarding, Gilding, and Finishing. Also, the Art of Marbling Book-edges and Paper. By JAMES B. NICHOLSON. Illustrated. 12mo., cloth. $2.25

NICHOLSON.—The Carpenter's New Guide:
A Complete Book of Lines for Carpenters and Joiners. By PETER NICHOLSON. The whole carefully and thoroughly revised by H. K. DAVIS, and containing numerous new and improved and original Designs for Roofs, Domes, etc. By SAMUEL SLOAN, Architect. Illustrated by 80 plates. 4to. . . . .

NORRIS.—A Hand-book for Locomotive Engineers and Machinists:
Comprising the Proportions and Calculations for Constructing Locomotives; Manner of Setting Valves; Tables of Squares, Cubes, Areas, etc., etc. By SEPTIMUS NORRIS, Civil and Mechanical Engineer. New edition. Illustrated. 12mo., cloth. $1.50

NYSTROM.—On Technological Education, and the Construction of Ships and Screw Propellers:
For Naval and Marine Engineers. By JOHN W. NYSTROM, late Acting Chief Engineer, U. S. N. Second edition, revised with additional matter. Illustrated by seven engravings. 12mo. . . $1.50

O'NEILL.—A Dictionary of Dyeing and Calico Printing:
Containing a brief account of all the Substances and Processes in use in the Art of Dyeing and Printing Textile Fabrics; with Practical Receipts and Scientific Information. By CHARLES O'NEILL, Analytical Chemist; Fellow of the Chemical Society of London; Member of the Literary and Philosophical Society of Manchester; Author of "Chemistry of Calico Printing and Dyeing." To which is added an Essay on Coal Tar Colors and their application to Dyeing and Calico Printing. By A. A. FESQUET, Chemist and Engineer. With an Appendix on Dyeing and Calico Printing, as shown at the Universal Exposition, Paris, 1867. In one volume, 8vo., 491 pages. . $5.00

ORTON.—Underground Treasures:
How and Where to Find Them. A Key for the Ready Determination of all the Useful Minerals within the United States. By JAMES ORTON, A. M. Illustrated, 12mo. $1.50

OSBORN.—American Mines and Mining:
Theoretically and Practically Considered. By Prof. H. S. OSBORN. Illustrated by numerous engravings. 8vo. (*In preparation.*)

OSBORN.—The Metallurgy of Iron and Steel:
Theoretical and Practical in all its Branches; with special reference to American Materials and Processes. By H. S. OSBORN, LL. D., Professor of Mining and Metallurgy in Lafayette College, Easton, Pennsylvania. Illustrated by numerous large folding plates and wood-engravings. 8vo. $15.00

OVERMAN.—The Manufacture of Steel:
Containing the Practice and Principles of Working and Making Steel. A Handbook for Blacksmiths and Workers in Steel and Iron, Wagon Makers, Die Sinkers, Cutlers, and Manufacturers of Files and Hardware, of Steel and Iron, and for Men of Science and Art. By FREDERICK OVERMAN, Mining Engineer, Author of the "Manufacture of Iron," etc. A new, enlarged, and revised Edition. By A. A. FESQUET, Chemist and Engineer. $1.50

OVERMAN.—The Moulder and Founder's Pocket Guide:
A Treatise on Moulding and Founding in Green-sand, Dry-sand, Loam, and Cement; the Moulding of Machine Frames, Mill-gear, Hollowware, Ornaments, Trinkets, Bells, and Statues; Description of Moulds for Iron, Bronze, Brass, and other Metals; Plaster of Paris, Sulphur, Wax, and other articles commonly used in Casting; the Construction of Melting Furnaces, the Melting and Founding of Metals; the Composition of Alloys and their Nature. With an Appendix containing Receipts for Alloys, Bronze, Varnishes and Colors for Castings; also, Tables on the Strength and other qualities of Cast Metals. By FREDERICK OVERMAN, Mining Engineer, Author of "The Manufacture of Iron." With 42 Illustrations. 12mo. $2.00

Painter, Gilder, and Varnisher's Companion:
Containing Rules and Regulations in everything relating to the Arts of Painting, Gilding, Varnishing, Glass-Staining, Graining, Marbling, Sign-Writing, Gilding on Glass, and Coach Painting and Varnishing; Tests for the Detection of Adulterations in Oils, Colors, etc.; and a Statement of the Diseases to which Painters are peculiarly liable, with the Simplest and Best Remedies. Sixteenth Edition. Revised, with an Appendix. Containing Colors and Coloring—Theoretical and Practical. Comprising descriptions of a great variety of Additional Pigments, their Qualities and Uses, to which are added, Dryers, and Modes and Operations of Painting, etc. Together with Chevreul's Principles of Harmony and Contrast of Colors. 12mo., cloth. $1.50

PALLETT.—The Miller's, Millwright's, and Engineer's Guide.
By Henry Pallett. Illustrated. In one volume, 12mo. $3.00

PERCY.—The Manufacture of Russian Sheet-Iron.
By John Percy, M.D., F.R.S., Lecturer on Metallurgy at the Royal School of Mines, and to The Advanced Class of Artillery Officers at the Royal Artillery Institution, Woolwich; Author of "Metallurgy." With Illustrations. 8vo., paper. . . . . . .   50 cts.

PERKINS.—Gas and Ventilation.
Practical Treatise on Gas and Ventilation. With Special Relation to Illuminating, Heating, and Cooking by Gas. Including Scientific Helps to Engineer-students and others. With Illustrated Diagrams. By E. E. Perkins. 12mo., cloth. $1.25

PERKINS and STOWE.—A New Guide to the Sheet-iron and Boiler Plate Roller:
Containing a Series of Tables showing the Weight of Slabs and Piles to produce Boiler Plates, and of the Weight of Piles and the Sizes of Bars to produce Sheet-iron; the Thickness of the Bar Gauge in decimals; the Weight per foot, and the Thickness on the Bar or Wire Gauge of the fractional parts of an inch; the Weight per sheet, and the Thickness on the Wire Gauge of Sheet-iron of various dimensions to weigh 112 lbs. per bundle; and the conversion of Short Weight into Long Weight, and Long Weight into Short. Estimated and collected by G. H. Perkins and J. G. Stowe. $2.50

PHILLIPS and DARLINGTON.—Records of Mining and Metallurgy;
Or Facts and Memoranda for the use of the Mine Agent and Smelter. By J. Arthur Phillips, Mining Engineer, Graduate of the Imperial School of Mines, France, etc., and John Darlington. Illustrated by numerous engravings. In one volume, 12mo.   . . $1.50

PROTEAUX.—Practical Guide for the Manufacture of Paper and Boards.
By A. Proteaux, Civil Engineer, and Graduate of the School of Arts and Manufactures, and Director of Thiers' Paper Mill, Puy-de-Dôme. With additions, by L. S. Le Normand. Translated from the French, with Notes, by Horatio Paine, A. B., M. D. To which is added a Chapter on the Manufacture of Paper from Wood in the United States, by Henry T. Brown, of the "American Artisan." Illustrated by six plates, containing Drawings of Raw Materials, Machinery, Plans of Paper-Mills, etc., etc. 8vo. $10.00

REGNAULT.—Elements of Chemistry.
By M. V. Regnault. Translated from the French by T. Forrest Betton, M. D., and edited, with Notes, by James C. Booth, Melter and Refiner U. S. Mint, and Wm. L. Faber, Metallurgist and Mining Engineer. Illustrated by nearly 700 wood engravings. Comprising nearly 1500 pages. In two volumes, 8vo., cloth. . . . $7.50

REID.—A Practical Treatise on the Manufacture of Portland Cement:
By HENRY REID, C. E. To which is added a Translation of M. A. Lipowitz's Work, describing a New Method adopted in Germany for Manufacturing that Cement, by W. F. REID. Illustrated by plates and wood engravings. 8vo. $5.00

RIFFAULT, VERGNAUD, and TOUSSAINT.—A Practical Treatise on the Manufacture of Varnishes.
By M M. RIFFAULT, VERGNAUD, and TOUSSAINT. Revised and Edited by M. F. MALEPEYRE and Dr. EMIL WINCKLER. Illustrated. In one volume, 8vo. (*In preparation.*)

RIFFAULT, VERGNAUD, and TOUSSAINT.—A Practical Treatise on the Manufacture of Colors for Painting:
Containing the best Formulæ and the Processes the Newest and in most General Use. By M M. RIFFAULT, VERGNAUD, and TOUSSAINT. Revised and Edited by M. F. MALEPEYRE and Dr. EMIL WINCKLER. Translated from the French by A. A. FESQUET, Chemist and Engineer. Illustrated by Engravings. In one volume, 650 pages, 8vo. **$7.50**

ROBINSON.—Explosions of Steam Boilers:
How they are Caused, and how they may be Prevented. By J. R. ROBINSON, Steam Engineer. 12mo. $1.25

ROPER.—A Catechism of High Pressure or Non-Condensing Steam-Engines:
Including the Modelling, Constructing, Running, and Management of Steam Engines and Steam Boilers. With Illustrations. By STEPHEN ROPER, Engineer. Full bound tucks . . . $2.00

ROSELEUR.—Galvanoplastic Manipulations:
A Practical Guide for the Gold and Silver Electro-plater and the Galvanoplastic Operator. Translated from the French of ALFRED ROSELEUR, Chemist, Professor of the Galvanoplastic Art, Manufacturer of Chemicals, Gold and Silver Electro-plater. By A. A. FESQUET, Chemist and Engineer. Illustrated by over 127 Engravings on wood. 8vo., 495 pages. $6.00

☞ *This Treatise is the fullest and by far the best on this subject ever published in the United States.*

SCHINZ.—Researches on the Action of the Blast Furnace.
By CHARLES SCHINZ. Translated from the German with the special permission of the Author by WILLIAM H. MAW and MORITZ MULLER. With an Appendix written by the Author expressly for this edition. Illustrated by seven plates, containing 28 figures. In one volume, 12mo. $4.00

SHAW.—Civil Architecture:
Being a Complete Theoretical and Practical System of Building, containing the Fundamental Principles of the Art. By EDWARD SHAW, Architect. To which is added a Treatise on Gothic Architecture, etc. By THOMAS W. SILLOWAY and GEORGE M. HARDING, Architects. The whole illustrated by One Hundred and Two quarto plates finely engraved on copper. Eleventh Edition. 4to., cloth. . $10.00

SHUNK.—A Practical Treatise on Railway Curves and Location, for Young Engineers.
By WILLIAM F. SHUNK, Civil Engineer. 12mo. . . $2.00

SLOAN.—American Houses:
A variety of Original Designs for Rural Buildings. Illustrated by 26 colored Engravings, with Descriptive References. By SAMUEL SLOAN, Architect, author of the "Model Architect," etc., etc. 8vo. $1.50

SMEATON.—Builder's Pocket Companion:
Containing the Elements of Building, Surveying, and Architecture; with Practical Rules and Instructions connected with the subject. By A. C. SMEATON, Civil Engineer, etc. In one volume, 12mo. $1.50

SMITH.—A Manual of Political Economy.
By E. PESHINE SMITH. A new Edition, to which is added a full Index. 12mo., cloth. $1.25

SMITH.—Parks and Pleasure Grounds:
Or Practical Notes on Country Residences, Villas, Public Parks, and Gardens. By CHARLES H. J. SMITH, Landscape Gardener and Garden Architect, etc., etc. 12mo. $2.25

SMITH.—The Dyer's Instructor:
Comprising Practical Instructions in the Art of Dyeing Silk, Cotton, Wool, and Worsted, and Woollen Goods: containing nearly 800 Receipts. To which is added a Treatise on the Art of Padding; and the Printing of Silk Warps, Skeins, and Handkerchiefs, and the various Mordants and Colors for the different styles of such work. By DAVID SMITH, Pattern Dyer. 12mo., cloth. . . . $3.00

SMITH.—The Dyer's Instructor:
Comprising Practical Instructions in the Art of Dyeing Silk, Cotton, Wool, and Worsted and Woollen Goods. Third Edition, with many additional Receipts for Dyeing the New Alkaline Blues and Night Greens, *with Dyed Patterns affixed.* 12mo., pp. 394, cloth. . $10.50

STEWART.—The American System.
Speeches on the Tariff Question, and on Internal Improvements, principally delivered in the House of Representatives of the United States. By ANDREW STEWART, late M. C. from Pennsylvania. With a Portrait, and a Biographical Sketch. In one volume, 8vo., 407 pages. $3.00

STOKES.—Cabinet-maker's and Upholsterer's Companion:
Comprising the Rudiments and Principles of Cabinet-making and Upholstery, with Familiar Instructions, illustrated by Examples for attaining a Proficiency in the Art of Drawing, as applicable to Cabinet-work; the Processes of Veneering, Inlaying, and Buhl-work; the Art of Dyeing and Staining Wood, Bone, Tortoise Shell, etc. Directions for Lackering, Japanning, and Varnishing; to make French Polish; to prepare the Best Glues, Cements, and Compositions, and a number of Receipts particularly useful for workmen generally. By J. STOKES. In one volume, 12mo. With Illustrations. . $1.25

Strength and other Properties of Metals:
Reports of Experiments on the Strength and other Properties of Metals for Cannon. With a Description of the Machines for testing Metals, and of the Classification of Cannon in service. By Officers of the Ordnance Department U. S. Army. By authority of the Secretary of War. Illustrated by 25 large steel plates. In one volume, 4to. . $10.00

SULLIVAN.—Protection to Native Industry.
By Sir EDWARD SULLIVAN, Baronet, author of "Ten Chapters on Social Reforms." In one volume, 8vo. $1.50

Tables Showing the Weight of Round, Square, and Flat Bar Iron, Steel, etc.,
By Measurement. Cloth. 63

TAYLOR.—Statistics of Coal:
Including Mineral Bituminous Substances employed in Arts and Manufactures; with their Geographical, Geological, and Commercial Distribution and Amount of Production and Consumption on the American Continent. With Incidental Statistics of the Iron Manufacture. By R. C. TAYLOR. Second edition, revised by S. S. HALDEMAN. Illustrated by five Maps and many wood engravings. 8vo., cloth. $10.00

TEMPLETON.—The Practical Examinator on Steam and the Steam-Engine:
With Instructive References relative thereto, arranged for the Use of Engineers, Students, and others. By WM. TEMPLETON, Engineer. 12mo. $1.25

THOMAS.—The Modern Practice of Photography.
By R. W. THOMAS, F. C. S. 8vo., cloth. 75

THOMSON.—Freight Charges Calculator.
By ANDREW THOMSON, Freight Agent. 24mo. . . . $1.25

TURNING: Specimens of Fancy Turning Executed on the Hand or Foot Lathe:
With Geometric, Oval, and Eccentric Chucks, and Elliptical Cutting Frame. By an Amateur. Illustrated by 30 exquisite Photographs. 4to. $3.00

Turner's (The) Companion:
Containing Instructions in Concentric, Elliptic, and Eccentric Turning; also various Plates of Chucks, Tools, and Instruments; and Directions for using the Eccentric Cutter, Drill, Vertical Cutter, and Circular Rest; with Patterns and Instructions for working them. A new edition in one volume, 12mo. $1.50

URBIN.—BRULL.—A Practical Guide for Puddling Iron and Steel.
By ED. URBIN, Engineer of Arts and Manufactures. A Prize Essay read before the Association of Engineers, Graduate of the School of Mines, of Liege, Belgium, at the Meeting of 1865-6. To which is added A COMPARISON OF THE RESISTING PROPERTIES OF IRON AND STEEL. By A. BRULL. Translated from the French by A. A. FESQUET, Chemist and Engineer. In one volume, 8vo. $1.00

VAILE.—Galvanized Iron Cornice-Worker's Manual:
Containing Instructions in Laying out the Different Mitres, and Making Patterns for all kinds of Plain and Circular Work. Also, Tables of Weights, Areas and Circumferences of Circles, and other Matter calculated to Benefit the Trade. By CHARLES A. VAILE, Superintendent "Richmond Cornice Works," Richmond, Indiana. Illustrated by 21 Plates. In one volume, 4to. $5.00

VILLE.—The School of Chemical Manures:
Or, Elementary Principles in the Use of Fertilizing Agents. From the French of M. GEORGE VILLE, by A. A. FESQUET, Chemist and Engineer. With Illustrations. In one volume, 12 mo. . . $1.25

VOGDES.—The Architect's and Builder's Pocket Companion and Price Book:
Consisting of a Short but Comprehensive Epitome of Decimals, Duodecimals, Geometry and Mensuration; with Tables of U. S. Measures, Sizes, Weights, Strengths, etc., of Iron, Wood, Stone, and various other Materials, Quantities of Materials in Given Sizes, and Dimensions of Wood, Brick, and Stone; and a full and complete Bill of Prices for Carpenter's Work; also, Rules for Computing and Valuing Brick and Brick Work, Stone Work, Painting, Plastering, etc. By FRANK W. VOGDES, Architect. Illustrated. Full bound in pocket-book form. $2.00
Bound in cloth. 1.50

WARN.—The Sheet-Metal Worker's Instructor:
For Zinc, Sheet-Iron, Copper, and Tin-Plate Workers, etc. Containing a selection of Geometrical Problems; also, Practical and Simple Rules for describing the various Patterns required in the different branches of the above Trades. By REUBEN H. WARN, Practical Tin-plate Worker. To which is added an Appendix, containing Instructions for Boiler Making, Mensuration of Surfaces and Solids, Rules for Calculating the Weights of different Figures of Iron and Steel, Tables of the Weights of Iron, Steel, etc. Illustrated by 32 Plates and 37 Wood Engravings. 8vo. $3.00

WATSON.—A Manual of the Hand-Lathe:
Comprising Concise Directions for working Metals of all kinds, Ivory, Bone and Precious Woods; Dyeing, Coloring, and French Polishing; Inlaying by Veneers, and various methods practised to produce Elaborate work with Dispatch, and at Small Expense. By EGBERT P. WATSON, late of "The Scientific American," Author of "The Modern Practice of American Machinists and Engineers." Illustrated by 78 Engravings. $1.50

WATSON.—The Modern Practice of American Machinists and Engineers:
Including the Construction, Application, and Use of Drills, Lathe Tools, Cutters for Boring Cylinders, and Hollow Work Generally, with the most Economical Speed for the same; the Results verified by Actual Practice at the Lathe, the Vice, and on the Floor. Together with Workshop Management, Economy of Manufacture, the Steam-Engine, Boilers, Gears, Belting, etc., etc. By EGBERT P. WATSON, late of the "Scientific American." Illustrated by 86 Engravings. In one volume, 12mo. $2.50

WATSON.—The Theory and Practice of the Art of Weaving by Hand and Power:
With Calculations and Tables for the use of those connected with the Trade. By JOHN WATSON, Manufacturer and Practical Machine Maker. Illustrated by large Drawings of the best Power Looms. 8vo. $7.50

WEATHERLY.—Treatise on the Art of Boiling Sugar, Crystallizing, Lozenge-making, Comfits, Gum Goods.
12mo. $2.00

WILL.—Tables for Qualitative Chemical Analysis.
By Professor HEINRICH WILL, of Giessen, Germany. Seventh edition. Translated by CHARLES F. HIMES, Ph. D., Professor of Natural Science, Dickinson College, Carlisle, Pa. . . . $1.50

WILLIAMS.—On Heat and Steam:
Embracing New Views of Vaporization, Condensation, and Explosions. By CHARLES WYE WILLIAMS, A. I. C. E. Illustrated. 8vo. $3.50

WOHLER.—A Hand-Book of Mineral Analysis.
By F. WOHLER, Professor of Chemistry in the University of Göttingen. Edited by HENRY B. NASON, Professor of Chemistry in the Rensselaer Polytechnic Institute, Troy, New York. Illustrated. In one volume, 12mo. $3 00

WORSSAM.—On Mechanical Saws:
From the Transactions of the Society of Engineers, 1869. By S. W. WORSSAM, Jr. Illustrated by 18 large plates. 8vo. . . $5.00

RECENT ADDITIONS TO OUR LIST.

AUERBACH.—Anthracen: Its Constitution, Properties, Manufacture, and Derivatives, including Artificial Alizarin, Anthrapurpurin, with their applications in Dyeing and Printing.
By G. AUERBACH. Translated and edited by WM. CROOKES, F. R. S. 8vo. $5.00

BECKETT.—Treatise on Clocks, Watches and Bells.
By SIR EDMUND BECKETT, Bart. Illustrated. 12mo. . $1.75

BARLOW.—The History and Principles of Weaving, by Hand and by Power.
Several Hundred Illustrations. 8vo. $10.00

BOURNE.—Recent Improvements in the Steam Engine.
By JOHN BOURNE, C. E. Illustrated. 16mo. . . . $1.50

CLARK.—Fuel: Its Combustion and Economy.
By D. KINNEAR CLARK, C. E. 144 Engravings. 12mo. . $1.50

CRISTIANI.—Perfumery and Kindred Arts.
By R. S. CRISTIANI. 8vo. $5.00

COLLENS.—The Eden of Labor, or the Christian Utopia.
12mo. Paper, $1.00; Cloth, ... $1.25

CUPPER.—The Universal Stair Builder.
Illustrated by 29 plates. 4to. $2.50

COOLEY.—A Complete Practical Treatise on Perfumery.
By A. J. COOLEY. 12mo. $1.50

DAVIDSON.—A Practical Manual of House Painting, Graining, Marbling and Sign Writing:
With 9 Colored Illustrations of Woods and Marbles, and many Wood Engravings. 12mo. $3.00

EDWARDS.—A Catechism of the Marine Steam Engine.
By EMORY EDWARDS. Illustrated. 12mo. . . . $2.00

HASERICK.—The Secrets of the Art of Dyeing Wool, Cotton, and Linen:
Including Bleaching and Coloring Wool and Cotton Hosiery and Random Yarns. By E. C. HASERICK. Illustrated by 323 Dyed Patterns of the Yarns or Fabrics. 8vo. $25.00

HENRY.—The Early and Later History of Petroleum.
By J. T. HENRY. Illustrated. 8vo. $4.50

KELLOGG.—A New Monetary System.
By ED. KELLOGG. Fifth Edition. Edited by MARY KELLOGG PUTNAM. 12mo. Paper, $1.00; Cloth, ... $1.50

KEMLO.—Watch Repairer's Hand-Book.
Illustrated. 12mo. $1.25

MORRIS.—Easy Rules for the Measurement of Earthworks by means of the Prismoidal Formula.
By ELWOOD MORRIS, C. E. 8vo. $1.50

McCULLOCH.—Distillation, Brewing and Malting.
By J. C. McCULLOCH. 12mo. $1.00

NEVILLE.—Hydraulic Tables, Co-Efficients, and Formulæ for Finding the Discharge of Water from Orifices, Notches, Weirs, Pipes, and Rivers.
Illustrated. 12mo. $5.00

NICOLLS.—The Railway Builder.
A Hand-book for Estimating the Probable Cost of American Railway Construction and Equipment. By WM. J. NICOLLS, C. E. Pocket-book Form. $2.00

NORMANDY.—The Commercial Hand-book of Chemical Analysis.
By H. M. NOAD, Ph. D. 12mo. $5.00

PROCTOR.—A Pocket-Book of Useful Tables and Formulæ for Marine Engineers.
By FRANK PROCTOR. Pocket-book Form. . . . $2.00

ROSE.—The Complete Practical Machinist:
Embracing Lathe Work, Vise Work, Drills and Drilling, Taps and Dies, Hardening and Tempering, the Making and Use of Tools, etc., etc. By JOSHUA ROSE. 130 Illustrations. 12mo. . . $2.50

SLOAN.—Homestead Architecture.
By SAMUEL SLOAN, Architect. 200 Engravings. 8vo. . $3.50

SYME.—Outlines of an Industrial Science.
By DAVID SYME. 12mo. $2.00

WARE.—The Coachmaker's Illustrated Hand-Book.
Fully Illustrated. 8vo. $3.00

WIGHTWICK.—Hints to Young Architects.
Numerous Wood Cuts. 12mo. $2.00

WILSON.—First Principles of Political Economy.
12mo. $1.50

WILSON.—A Treatise on Steam Boilers, their Strength, Construction, and Economical Working.
By ROBT. WILSON. Illustrated. 12mo. $2.50

www.ingramcontent.com/pod-product-compliance
Lightning Source LLC
Chambersburg PA
CBHW031742230426
43669CB00007B/452